Sébastien Kidushi

Performance and potential of local plants

AF376122

Sébastien Kidushi

Performance and potential of local plants

in the agroforestry system of Ibi-village, Plateau des Batékés, Kinshasa/DRC. Case of edible caterpillar plants

ScienciaScripts

Imprint

Any brand names and product names mentioned in this book are subject to trademark, brand or patent protection and are trademarks or registered trademarks of their respective holders. The use of brand names, product names, common names, trade names, product descriptions etc. even without a particular marking in this work is in no way to be construed to mean that such names may be regarded as unrestricted in respect of trademark and brand protection legislation and could thus be used by anyone.

Cover image: www.ingimage.com

This book is a translation from the original published under ISBN 978-620-3-42967-1.

Publisher:
Sciencia Scripts
is a trademark of
Dodo Books Indian Ocean Ltd. and OmniScriptum S.R.L Publishing group
Str. Armeneasca 28/1, office 1, Chisinau MD-2012, Republic of Moldova, Europe
Printed at: see last page
ISBN: 978-620-5-36933-3

Copyright © Sébastien Kidushi
Copyright © 2022 Dodo Books Indian Ocean Ltd. and OmniScriptum S.R.L Publishing group

Contents

EPIGRAPH

<< what the caterpillar calls death, the butterfly calls rebirth "

Violette Lebon

DEDICATION

To my parents, Kidushi Kapay Romain and Kanikani Awowo Tьёгезе,

To all the Kidushi, Kasay, Ingongo, Kanikani families,

To my aunts and uncles, Godelieve Kanikani, Guy Kanikani, Baudouin Kanikani, Micheline Kanikani, Virginie Kanikani and Georgine Kanikani,

To my brothers and sisters, Anicet Ingongo, Romulus Kidushi, Rooney Kidushi, Corel Ingongo, Maël Kasay, Alix Lejoly, Alex Kasay, Nodriche Kasay, Christelle Kasay, Grace Kasay, Lynda Ingongo, Diane Ingongo, Rolive Ingongo, Romaine Kidushi, Goffinete Lejoly

To the late Remus Kidushi, Martine Vumunene, Marie Masiwe, Joachim Kanikani and Sëbastien Kidushi.

To my friends Judith Mpembe, Esther Yoka, ,кш! ё Djedi, Cement Beloko, Frëdëric Ndege, Amina Ngala, Rodrigue Imwa, Chancelvie Ngyengo, Credo Mesongolo, Deo Mawete, Jërëmie Topesa, Sandra Muntazini, Sylvestre Frey, Aimëe Bole, Maria Masakala, Azarias Mwatoike and others,

ACKNOWLEDGEMENTS

At the end of this work, we would like to express our gratitude to all those who have helped us scientifically, materially and morally, from near or far, to carry out this work.

For this, we thank Almighty God first and foremost for His infinite goodness and mercy.

From a bacon райсиНёге, we express our deep thanks and appreciation to Professors Honore Belesi Katula and Jean Lejoly respectively director and supervisor, who, with tireless patience, scientific rigour and great availability, agreed to sponsor this work.

May the academic authorities, the academic, scientific, administrative, technical and working staff of the University of Kinshasa, find here the expression of our deep gratitude.

We are also indebted to Professor Jean Lejoly and his wife, Ms Micheline Kanikani, for their guidance and wise counsel.

We express, from bacon particulubre, the same feeling of gratitude to the Rëvërende Sreur Georgine Kanikani for her moral support.

To all those who contributed to the collection of data in the field, to the team of forestry engineers of lbi-village, we say thank you.

Our thanks also go to all our family. May all the Kidushi and Kanikani families, our brothers and sisters, cousins, nephews and nièces, uncles and aunts, friends and comrades find, through these lines, the expression of our gratitude.

May my friends and comrades: Azarias Mwatoike , Frederic Ndege , Chancelvie Ngyengo , Clement Beloko , Josue Djedi , Amina Ngala and Sandra Muntazini with whom we shared painful moments of fieldwork (thorns, bites of wasps and other harmful insects) and all those whose names do not appear on this list, find here the expression of our deep gratitude for the warmth of their company throughout this uncertain path that we have travelled together.

SUMMARY

A study on the performance and potential of edible caterpillar plants planted in lbi-village FAS was conducted in the Bateke plateau in the Democratic Republic of Congo. The main objective of this study was to assess the performance and potential of edible caterpillar plants planted in lbi-village agroforestry systems and to do this we used observation as a research method. In this study, 35 species of edible caterpillar hosts were recorded, divided into 16 families and 7 orders. In terms of the ecological spectra of the floras, Mesophanerophytes and sarcochores are predominant. On the phytogeographical level, the strong representativeness of the Guinean-Congolese element. Morphologically, the majority of species are trees. In this study, the species *Pentaclethra eetveldeana* is the most dominant. The most represented families are *Fabaceae* and *Euphorbiaceae*. In terms of height performance, 29 species are less successful (0.2 to 1.3 m/year), 5 species have an average performance (1.4 to 2.5 m/year) and only one species is the most successful in height (2.6 to 3.7 m/year). In terms of diameter performance, 25 species are the least efficient (from 0.4 to 1.9 cm/year), 6 species have a diameter performance (from 2.0 to 3.5 cm/year) and finally 4 species are the most efficient (from 3.6 to 5.1 cm/year). Zoologically, 3 species of edible caterpillars were identified in the Ibi region and were grouped into 2 families: *Notodontidae* and *Nymphalidae*. The harvesting period of these caterpillars is from February to December. Of these species, *Imbrasia ertli* is the most widely collected and 14 potential edible caterpillars can nest in our study area. The caterpillar plants planted in SAF Ibi have multiple uses.

Keywords: Performance, potential, trees, caterpillars, SAF, Bateke Plateau, Democratic Republic of Congo.

INTRODUCTION
1. Problematic

In the DRC, the degradation of forest ecosystems is increasingly observed around large cities, in some cases up to a radius of 150 kilometres (MALELE, 2003 ; MALAISSE *et al.*, 1985). Forest galleries   are disappearing due to
This is the only way to ensure the supply of wood energy, LFPs and NTFPs to these settlements.

The dense forest is subject to mostly uncontrolled exploitation, which can lead to environmental changes throughout the country (MALOBA, 2011). Some initiatives developed in the vicinity of Kinshasa (Mampu and Ibi-village projects) do not cover the wood and energy needs of this large city where the population was estimated at 10 million in 2009 (www.statistiques- mondi ales.com/ congo kinshasa.htm).

Savannahs are vëgëtal formations that cover land that is relatively easy to cultivate, and are generally characterised by soils with low organic and mineral reserves that ensure only a fragile equilibrium (YOKA et al., 2007). However, these savannahs are crossed annually by common fires without it being possible to establish a general rule as to their date (ACHOUNDONG et al., 2007).

In the DRC, these fires are early during the short dry season (mid-January and February) and late during the long dry season (May and September). They therefore present the only anthropogenic factor undoubtedly slowing down the progression of the forest, (DEFORESTA, 1990). This situation hinders the natural regeneration of the forest species that have colonised these savannahs in the course of their history and of which there are still less obvious traces on the Batëkë plateau.

In the savannahs of the Bateke Plateau, the main types of human activity practised there are bush farming, the exploitation of wood in various forms (firewood, charcoal, timber for construction etc.), gathering and hunting. All these activities have become unsuccessful. The sandy soil has rapidly become impoverished, the destruction of animal nests has rarëйë the game, gathering is now only about a few fruits *of Anisophyllea quangensis* and *Landolphia lanceolata.*

Faced with the threats that these anthropogenic activities pose to these ëcosystëmes, a growing concern is developing around the probiems of biodiversity loss and degradation (IYONGO et al., 2012).

Insects (Termites, Grasshoppers, Crickets, Coleopteran larvae, ...) have contributed significantly to the diet of many populations for a very long time. They are very popular among sub-Saharan Africans (Democratic Republic of Congo, Central African Republic, Cameroon, Republic of Congo ...); among North American Indians, in Mexico as well as in many Asian (China, Japan) and African (Botswana, Zimbabwe) cultures. Nowadays, insect dishes can also be found on restaurant menus in various European countries such as France and Belgium (FAO, NTFP, 2004).

In the last decade, recognition of the role of edible non-timber forest products (NTFPs) in food security, such as game, fruits, local mushrooms, has increased significantly. Despite this, the dietary potential of edible insects is poorly known, although several studies have shown that insects make an important contribution to rural and urban livelihoods.

In many cultures, insects are consumed as a daily supplement, a delicacy, an occasional meal or a substitute during food shortages, droughts, floods, wars, etc. Edible insects should be considered as a more important potential alternative in efforts to improve food security and alleviate poverty in sub-Saharan Africa, in general, and in the Democratic Republic of Congo, in particular (DE FOLIART, 1992).

Researchers and others working in development organisations know little about how to optimise the potential of insects, especially those that come from forests and are edible. Less is known about the links between forest management and insect populations, the impacts of insect collection on forest and biodiversity in general, and the dietary patterns in local diets related to changes in the availability of other protein sources, particularly game.

The richness and diversity of edible caterpillar host plants present, both in production forests and in galleries, wooded and low herbaceous formations, and in plantations, constitute an asset whose sustainable management must stimulate its perpetuation. The collection and exchange of edible caterpillars remains an informal activity, and their commercialisation provides an interesting but underestimated economic complement (N'GASSE, 2003).

Several species of edible caterpillars are collected periodically throughout the country and are present in most Congolese markets. According to estimates by MONZAMBE (2002), more than 85% of the population consumes them (fresh, dry or boiled).

The need to carry out inventories and collect data on the dynamics of caterpillars, the interactions of human populations with this specific environment, and their socio-economic impacts, will make it possible to understand some of the problems involved in exploiting this resource and to formulate proposals likely to improve the livelihoods and increase the incomes of the Congolese.

Thus, studies have been oriented towards the planting of multiple plants, including edible caterpillars in agroforestry systems to reduce anthropogenic pressures on natural plant formations, allow a return of caterpillars and promote the restoration of natural ecosystems. Indeed, in order to adapt to new constraints, and to ensure a possible return of edible caterpillars and other non-wood forest products, the planting of edible caterpillar plants in agroforestry systems is becoming increasingly important.

Based on the different roles that plants play and the way they are exploited and used by communities, especially the people of the Bateke Plateau, we have highlighted some research questions to understand the performance and potential of different edible caterpillar plants planted in the ibi-village agroforestry system (FAS). Specifically, the following questions arise:

1.   can we find significant floristic diversity in edible caterpillar plants planted in the SAF of lbi-village?

2.   what is the growth performance of these species in height and diameter compared to other species planted in the region?

3.   What is the potential of these plants compared to other species planted in the Ibi-village FAS?

4.   what is the most common species of caterpillar collected and sold in the Ibi-Village area? what is the harvesting period for these caterpillars?

5.   What are the potential caterpillars that will be able to nest on the plants planted in SAF in Ibi-village?

In this field, a lot of previous work has been done by different authors.

The study of insects, and in particular caterpillars, has already been the subject of several scientific publications since the colonial period. Some of the works that have already been carried out in this field are the following:

-       **HEYMANS** and **EVRARD** (1970) contributed to the study of the food composition of edible insects in the Katanga province (DRC).

-       **LELEUP** and **DAEMS** (1969) contributed to the study of the Kwango food caterpillars, the causes of their scarcity and the measures recommended to remedy them (DRC).

-       **BOLA** (1973) studied caterpillars and their use in food among the Mongo in

Mbandaka, Equateur Province (DRC).

- **MALAISSE** and **PARENT** (1980) inventoried the edible caterpillars of Shaba meridional (zaire).

- **LATHAM** (2000) studied and analysed edible caterpillars and their food plants in Bas-Congo province (DRC).

- **MONZAMBE** (2002) studied the exploitation of caterpillars and other edible larvae in the fight against food insecurity and poverty in the Democratic Republic of Congo (DRC).

- **MOUSSA** (2002) studied the edible caterpillars of Congo: food interest and marketing channels.

- **BALINGA** (2003) analysed the contribution of caterpillars and edible larvae in the forest zone of Cameroon.

- **N'GASSE** (2003) analysed the contribution of edible caterpillars/larvae to the reduction of food insecurity in the Central African Republic (CAR).

- **MASSEGIUM** and **ANTONINI** (1938) listed 21 vernacular names of edible caterpillars in the Sangba forest (now Mambere kadeii, CAR).

- **OKANGOLA** (2007) contributed to the biological and ëcological study of edible caterpillars in the Kisangani region. Case of the Yoko reserve (Ubundu, DRC).

In the Bateke plateau around Ibi, the study of edible caterpillars and their host plants seems, from our literature search, not to have been sëeriously tubordeen

As such, we believe it is useful to fill this gap by conducting a study on the performance and potential of edible caterpillar plants planted in the lbi-village agroforestry system.

2. Assumptions

To answer the questions raised above, we formulate the hypotheses that :

- The Ibi region, in general, and Musiabikui, in particular, has a great diversity of caterpillar plants that are planted there;

- these planted caterpillar species show a diversity of growth performance (less, medium and strong) in diameter and height.

- edible caterpillar plants planted in SAF have multiple potentialities.

- *Imbrasia ertli* is the most collected species in the study area.

- 16 potential edible caterpillars will nest on caterpillar plants planted in SAF of Ibi-village.

3. Objectives

The general objective of this work is to evaluate the performance and potential of caterpillar plants planted in the Ibi-village FAS.

Specifically, it is about :

-        Identify the different species of edible caterpillar plants planted in SAF in Ibi-village.

-        to clarify the growth performance of caterpillar plants grown in the
forest agrosystems of Musiabikui/Ibi.

-        clarify the potential of caterpillar plants in the Ibi-village FAS.

-        to identify the species of caterpillars most commonly collected and sold in the Ibi
area.

-        find potential caterpillars that may nest on this vegetation.

4. Interest of the subject

As a result of this study, we have gained a better understanding of the edible caterpillars and their host plants.

In addition, the results of this work will provide baseline data for further studies in the zoological, botanical, nutritional, economic and other fields.

They will also provide information that can help preserve the host plant species of edible caterpillars.

The data collected in this work will be used by the populations of the region to integrate this field into their sustainable development projects within the framework of agroforestry and to valorise the consumption of caterpillars.

5. Delimitation of work in time and space.

The work is limited both in space and time. In time this work goes from June 2020 to July 2021 and in space it is carried out in the city province of Kinshasa, precisely in the urban-rural commune of Maluku, Mbankana district, Ibi village on the Bateke plateau.

6. Labour Subdivision

In addition to the introductory part and the conclusion, this work consists of three chapters, the first chapter deals with generalities, the second chapter deals with the study environment, materials and methods and the third chapter deals with the presentation of the results and the discussion.

CHAPTER I

GENERAL

In this chapter, we discuss the gënëralitës on the key concepts of our work, especially on caterpillars and the agroforestry system.

1.1. General information on caterpillars

a)   Definition

Caterpillars are butterfly larvae with very different colours and shapes, but their basic structure is much the same. Butterflies belong to the Order Lëpidoptëres, in the class Insects, in the phylum Arthropoda and in the Animal kingdom (CARTER *et al,* 1998).

b)   Morphology

The caterpillar has a relatively simple organisational plan. It has a head followed by 13 rings, the first 3 of which form the thorax and the others the abdomen (fig. 1).

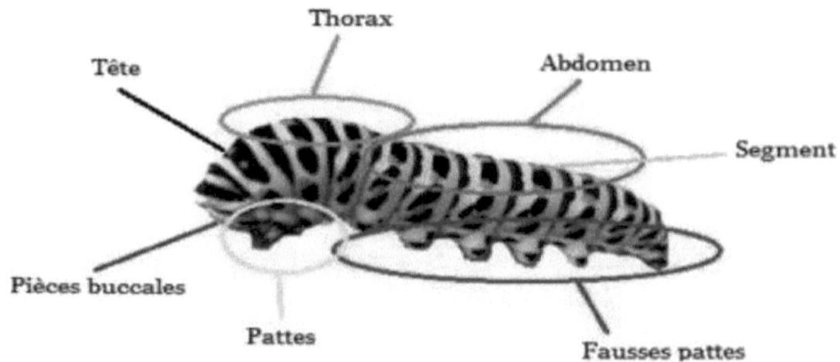

Figure 1: *The morphology of the caterpillar.*

b.1. Head

The head is sclerified (cephalic capsule), bearing sensory setae, various appendages and lined with grinder-type mouthparts that are made for biting and chewing. The oral pièces consist of a pair of mandibles, a pair of maxillae, a labium (lower yeast) and a labre (upper yeast). The labium has two palps (sensory elements) and a spinneret, the apical opening of which corresponds to the opening of the sericid glands.

The liquid ë put in by the latter solidifies on contact with the air and a silk thread is thus produced, which is very resistant despite its thinness.

Silk has many uses:

-        it is used to make cocoons;

-        it allows the collection of leaves or other objects for the construction of larval shelters;

-        it can form mats that caterpillars cling to (especially during moulting) or suspension threads that they use to drop down (in case of danger).

On either side of the labium are the maxillary palps, which appreciate the taste of food while directing it towards the mouth. Near the mandibles are antennae (sensory organs).

The head capsule consists essentially of an anterior frontal area and two large lateral pieces called epicrania (the latter form the vertex). The ventral part of each epicranium usually shows 6 rudimentary and isolated eyes called stemmata.

b.2. Thorax

The thorax is made up of 3 segments (prothorax, mesothorax and metathorax) each bearing a pair of jointed legs with a single (odd) claw.

The legs are more involved in grasping food than in locomotion. The prothorax is often distinguished from the other two segments by the presence of a more or less sclerified dorsal piece: the pronotal plate.

b.3. Abdomen

The abdomen has 10 segments, 5 of which have a pair of false legs with pincers (hooks), which are used to attach and move the caterpillar.

These false legs are located on the 3rd to 6th and 10th segments with false anal legs forming the anal patch.

c) Life cycle

Butterflies are endoptërygous (holomëtabolic) insects, i.e. insects that undergo a competitive transformation or mëtamorphosis during their development.

These insects have 4 distinct ëtages: egg, larva (or caterpillar different from butterfly in appearance and life form), pupa (or chrysalis) and imago (adult or butterfly). (SMART, 1981).

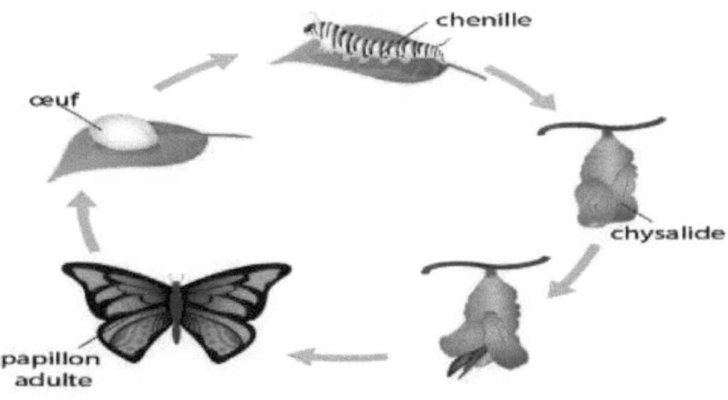

Figure 2: *life cycle of the caterpillar*

- L'reuf :

The eggs are deposited by the females on a plant at specific points (leaves or flowers), glued by a sticky secretion. The female attaches herself to the top of the leaf and folds her abdomen to place it under the underside.

Most females install them on the underside of a leaf to protect them from rain, sun and, to some extent, predators. The arrangement and shape of the eggs varies from species to species. They may be arranged in regular bands around the twigs or in a string in a rounded, oval or flattened shape.

-   Caterpillar :

This is the moth larva obtained after hatching. The consumption of leaves is the main activity of the larva and determines its fate. During its growth, the larva eventually fills the entire epidermis, which becomes distended.

The size increases and is accompanied by successive changes in the skin. A new, larger ëpidermis forms under the first.

The new skin grows rapidly before the skin hardens. This change of skin is called moulting and takes place 4 to 5 times before the end of growth.

The periods between moults are called instars. In butterflies, moulting is controlled by hormones. However, the environment and the availability of food can affect the duration of the instars.

-      **Chrysalis :**

The chrysalis is the stage that marks the end of the larval life by a final moult. When the caterpillars have reached the end of their growth, they leave their host plants, start wandering to find a suitable place to pupate, sometimes far from their feeder plants, and attach themselves to the substrates, places to weave the silk mats that will serve as support.

They build cocoons (shelters) which are simple leaves joined by a few threads of silk, sometimes silky pouches made of various debris.

When pupation takes place in the soil, the chrysalis is usually found inside an earthy, silk-lined cocoon. The cocoon keeps the chrysalis in satisfactory hygrometric conditions and helps to protect it from parasites and predators.

During this stage, the chrysalis is immobile, does not eat or drink, because the mouth and anus are blocked. The only functioning openings are the stigmata which allow respiratory exchange. The legs and antennae, glued to the body, cannot move. This is a state in which important internal changes take place, leading to the appearance of the butterfly's organs.

- Imago :

This is the butterfly; it represents the stage where reproduction and dispersal of the species takes place. The duration of each phase of the cycle varies according to the species considered and even the climatic conditions in which it flies. Generally, eggs hatch after one to four weeks.

Similarly, some caterpillars develop in a few weeks, others take at least several months (sometimes 2 years) to reach their maximum size: these are the species that consume wood or nutrient-poor material.

The duration of the pupal stage, which is the longest, is variable and ranges from a few days to several years. Some open quickly, others only emerge after the bad season; in some cases the emergence period extends over several months.

Most imagos live only a few weeks, others up to ten months. This high variability in the life cycle of butterflies has allowed them to adapt to a wide range of feeding conditions (CARTER AND HARGREAVES, 1981).

1.2. Some concepts on vegetation.

1.2.1. **Drill**

The term forest has several definitions according to the authors. We will retain two definitions including (1) the one communicated to the United Nations Framework Convention on Climate Change (UNFCCC) and (2) the one developed in the DRC Forestry Code (2002).

The forest is thus defined as, respectively, a piece of land that has :

- either a minimum tree cover of 30%; for a minimum area of 0.5 hectares; with a minimum tree height of 3 metres.
- or covered by a vegetation formation based on trees or shrubs capable of providing forest products, sheltering wildlife and having a direct or indirect effect on the soil, climate or water regime.

Both definitions can be summarised according to BELESI (2020): a forest is a plant formation dominated by trees and shrubs, sparse of herbs with high humidity in the undergrowth.

In the DRC, despite its history, the forest that ëtait a йег! ё in the central basin is in the process of joining others such as the Amazon (Brazil) and Indonesia which have been under heavy anthropogenic pressure. Hence the global strategy to restore them through afforestation and reforestation.

1.2.2. **Reforestation**

Reforestation is defined as the direct human-induced conversion of non-forest land to forest land by planting, seeding and/or human promotion of natural seeding on land that had previously been forested but has been converted to non-forest land (IPCC, 2003).

1.2.3. **Deforestation**

The natural or anthropogenic process of transforming a forest into non-forest land, (IPCC, 2003).

1.2.4. **Afforestation**

Direct human conversion of land that has not been forested for at least 50 years into forested areas by means of planting, seeding and/or encouraging the use of natural seed resources, (IPCC, 2003).

1.2.5. **Plant**

Plants are multicellular beings at the base of the food chain. They form one of the subdivisions (or kingdom) of eukaryotes. They are, together with other plants, the object of study in botany (HAECKEL, 1866).

1.2.6. **Tree**

The tree is a large woody plant, the thickened stem forms the trunk or shaft. It supports a set of more or less branched branches called the houppier or crown (BELESI, 2015). For PARENT (1990), the tree is any woody vëgëtal of high height possessing a trunk that branches only from a certain height.

1.2.7. **Shrub**

The shrub is a woody plant of small size 4 to 6 m (BELESI, 2015). For PARENT (1990) the shrub is a woody vëgëtal with a total height infëior to 5 тëйез.

1.2.8. **Arboretum**

An arboretum is a botanical garden spëcialisë, generally conceived as a landscape space. It represents numerous species of trees or woody species in the form of collections, most often thematic (INERA).

1.2.9. **Habitat**

It is the place where an organism lives and meets in a usual way. For example, the bamboo forest in the mountainous regions of eastern DRC is the natural habitat of mountain gorillas. Similarly, the Guinean-Congolese grassy formations developed on the soils of former termiticres is the habitat of fungi called *Termitomyces striatus* (BELESI, 2020).

1.2.10. **Conservation**

Conservation is an approach that takes into account the long-term variability of ecosystems in resource and environmental management projects. In the Anglo-Saxon sense of the term, it is a protection that does not prohibit human intervention in natural processes. It is a philosophy of environmental management that does not lead to its waste or ëexhaustion (BELESI, 2017).

1.2.11. **Biological diversity**

It is the variability among living organisms from all sources including, inter alia, terrestrial, marine and other aquatic ecosystems and the ecological complexes of which they are part; this includes diversity within and between species as well as between ecosystems, according to the Convention on Biological Diversity (1992).

1.2.12. **Biological resources**

Are genetic resources, organisms or parts thereof, populations, or any other biotic component of ecosystems with actual or potential use or value for humanity, according to the Convention on Biological Diversity, (1992).

1.2.13. **Sustainable use**

Is the use of components of biological diversity in a way and at a rate that does not lead to their long-term decline, thereby safeguarding their potential to meet the needs and aspirations of present and future generations, according to the Convention on Biological Diversity (1992).

1.2.14. **Sustainable forestry**

is the sustainable use and management of forest ecosystems in such a way that it meets the needs of present generations without compromising the ability of future generations to use the forests, (RAVEN et al., 2008).

1.3. **Agroforestry**

Agroforestry refers to land use systems where trees or shrubs are grown in association with crops, pasture or livestock and where there are both ecological and economic interactions between the woody plants and other components, (YOUNG, 1995; ATANGANA *et al.*, 2014).

According to VON *et al*, (1982) quoted by BILOSO, (2008), the term agroforestry thus consists of the deliberate cultivation of woody perennial plants in association with agricultural crops or grazing, and these various types of production may coexist or succeed each other.

In any case, there is an ecological or economic interaction that can be positive or negative between woody vegetation and other components of the system.

MALDAGUE, (2010) quoted by BILOSO (2008) defines the term agroforestry as systems where three dimensions are involved:

- conservation (protection and rational use) of the biophysical environment ;
- multipurpose forest management (including dendro-energy and NTFPs);
- food production (plant, animal, including fish).

These different dimensions should be treated as components of a whole, of a system, and thus find their place in the framework of integrated rural development.

Thanks to the integration of trees in the rural system and the multiple functions they perform in the ecosystem, agroforestry systems, like the forest, contribute to the maintenance and even improvement of soil fertility. They thus have a positive effect that contributes to the sustainability of these production systems.

Thus, ATANGANA *et al*, (2014) redefine agroforestry as a collective name for any natural resource management system where perennial trees are spatially, temporally or spatio-temporally integrated on the same area with herbaceous or perennial crops of value (food, industrial, horticultural, forage, botanical, decorative, artistic) and/or livestock, terrestrial or aquatic organisms, with a view to diversifying and sustaining production for increased health and well-being of land users at all levels, depending on ecological, socio-economic, political and cultural circumstances.

Agroforestry encompasses many concepts and can be applied in very different ways. According to the World Agroforestry Centre, "trees are a source of products and services for rural and urban populations. While natural vegetation is gradually being removed for agricultural and other development purposes, the services provided by trees can be maintained by integrating them into agricultural systems", (TAFOKOU, 2014).

The relationship of agroforestry to soil conservation is influenced by climate, soil type and topography. The main components of agroforestry systems are woody plants, crops, pastures and livestock, and environmental factors (climate, soil, topography).

The domestication of agroforestry trees is a multidimensional process in which a progressive interaction between humans and plant resources takes place, WIERSUM, (1996) in DEGRANDE *et al*.

The domestication of local trees aims to increase, stabilise and diversify farmers' income sources and improve health care in rural areas, while encouraging the development of sustainable agroforestry practices.

In this context, domestication is therefore a natural resource management tool that requires the involvement of all stakeholders (DEGRANDE *et al.*, 2007). The domestication of trees is a much more recent phase than that of annual crops. According to the definition of LEAKEY & NEWTON, (1994) in BILOSO, (2008). According to WIERSUM (1996), tree domestication involves three stages:

- the shift from unregulated use of wild tree products to controlled exploitation ;

- protection and management in an agroforestry system or enriched environment ;

- propagation and cultivation of improved trees.

Thus, to domesticate a tree is to bring it from its wild state to a state where it undergoes a certain selection and management. It also means improving the tree so that it better corresponds not only to human needs and desires, but also to market expectations (LARSEN,

1993).

In this sense, domestication plays a very important role in poverty alleviation as it provides farmers with local forest products that are used daily for food, medical care and construction. It also contributes to the diversification of rural incomes and the development of communities (SCHRECKENBERG *et al.*, 2006).

1.3.1. Different forms of agroforestry

There are several forms of agroforestry which can be summarised as follows:

- agricultural plots planted with fruit or fodder trees ;
- grazing around wood-producing trees;
- line of windbreak trees ;
- crops in corridors alternating with alleys and woody hedges;
- Bush fallow ;
- forest plantation with annual undergrowth.

1 3.2. Benefits of trees in the agroforestry system

The bënëficial effects of the tree in the agroforestry system are multiple:

• The roots of the trees "pump" the mineral elements into the soil profile and bring them back to the surface through the vëgëtal debris (leaves, branches, etc.); this vëgëtal debris, at the soil level, is a source of energy for the pëdobiocënoses, resulting in an improvement of the soil structure and its enrichment in humus ;

• by the practice of felling, one obtains firewood;

• the effect of wood fragments on the organic matter and humus content of the soil is positive;

• trees associated with crops have a bio-static effect and favourably modify the meso climate (MALDAGUE, 2010).

1.4. Generalities about the savannah

The word "savanna" has a polysëmie dëveloppëe according to different geographical approaches. Thus, its definition varies according to biogeography, ecological conditions, climate, soils and geomorphology, (HIOL *et al.,* 2013).

Savannahs are tropical plant and animal formations dominated by heliophilous grass species of varying size and cover, with some woodland or scrub. Their biological rhythm is alternating (wet/dry season) (CAMARA, 2000).

According to DUVIGNAUD (1953), savannahs are purely grassy plant formations, i.e.

possessing, in addition to the grassy strata, a fairly loose shrub or tree layer whose constituents have no significant effect on the composition of the lower strata, AUBREVILLE (1957) and TROCHAIN (1957).

JACQUIN (2010) defines savannas as "open grassy formations composed of perennial and annual grasses". They are rather constituted by a very large proportion of tall grasses, especially gramineae, in tropical regions subject to a more or less long dry season. Among their characteristics, at least the following two criteria must be met to speak of savannas (MUHASHY *et al*, 2011):

- There are no trees, or if there are, they are less than 20 m high, the tops are very disjointed and the overall canopy is very open;

- presence of a gramiirean stratum facilitating the circulation of fire.

1.4.1. Origin of the savannahs

In humid tropical environments, dëforestation leads to vëgëtations herbacëes. In some situations, the savannah can be established permanently (AVENARD, 1969). In others, it is only a transient ëtate in the гедё^^^ Гогезйёге, (RAKOTOARIMANANA *et al*, 2008; GUNTHER *et al*. 2008).

The origin of the Central African savannas is controversial (KEAY, 1959; KLEIN, 2002; VANDE, 2004). It varies according to the authors: paleo-climatic, (BURNEY, 1997; HUMBERT, 1927; PERRIER, 1921), ëdaphic, through the action of fire, (KULL, 2004; BURNEY, 1996) or anthropic, (GADE, 1996).

The concept of savannah has been Гогдё since the second half of the XXëme siëcle with Anglo-Saxon and Francophone scientific reports and researchers from countries where this environment is present, (HIOL-HIOL et al., 2013). SCHNELL, (1971) quoted by FAO (http://www.fao.org/docrep/w4442f/w4442f08.htm, accessed on, 17.08.2015) proposes three explanatory scenarios that can serve as a repëres on the origin of savannas:

- **natural origin:** this conception envisages that savannas (mainly gramineous) have been established in environments that could not support abundant forest vegetation, due to poor soils or the existence of ancient limiting climates. In this approach, the savannah is the expression of a deficient and constrained environment. However, in the face of various considerations such as the development of certain savannas in humid zones "Esobe in the middle of the equatorial forest", the vegetal and animal characteristics of the population of various savannas, the pedological structures and the age of fires, the idea of natural savannas gives way, in some cases at least, to a delicate theory of savannas;

- **Delicate origin**: this conception is based on indications pleading for an ancient continuity of the savannas. They would have appeared during a drier period and would have been maintained thanks to the action of fires;

- **secondary origin:** in this evolutionary approach, a double origin of the savannas seems possible: either they derive from former dense hydrophilic formations replaced by less hydrophilic species (substitution savannas), or they succeed to dry tree formations (dry dense forests for example) and in this case, we speak of differentiated savannas in place. This secondarisation, mainly anthropogenic, is not easily proven and there is little evidence to confirm it. Today, most authors admit that they are maintained in their state by fire and that, consequently, they are non-climatic because, when they are cleared, they progress towards shrubland formations (SCHZART et al., 1996; ACHOUR et al., 2011; BADJI et al., 2013).

1.4.2. Savannah classification

Savannahs can be classified based on their physiognomic distinctiveness into three main groups: grassy savannahs, shrubby savannahs and shrubby savannahs.

- **Grassy savannahs**: these are those in which trees and shrubs are usually absent. This type of vegetation is found on all the itineraries observed on the Bateke plateau (MUHASHY *et al. Op.cit.*); the absence of trees or shrubs allows hëliophilous plants to develop.

- **Shrub savannahs**: are characterised by their physiognomy with two distinct strata: a grassy stratum composed of gramineae and little developed because it is dominated by a shrub stratum with woody elements between 1.80 m and 5 m high.

- **Arboreal savannas:** (or pre-forest) represent an intermediate formation between the forest and the shrub savannah. They are distinguished from the others in physiognomy and structure by the presence of trees. These are small in size but can exceed 7 m in height, covering more or less 20% of the habitat area (MUHASHY *et al*, 2011). All three types of savannah are present on the Bateke Plateau.

Several characteristics are common to these three savannah complexes: in all regions, savannahs coexist with forests and burn regularly in the dry season, they are threatened by desertification, climate change, unsustainable agricultural practices and environmental constraints, which leads to a decrease in their biological diversity and the appearance of invasive species, (HIOL *etal*; 2013).

CHAPTER 2

STUDY ENVIRONMENT, MATERIALS AND METHODS

In this chapter we will present our study environment, the material and the methods used for its realisation.

2.1. Study environment

In this section, we will first dëcribe the Batëkë plateau as a whole and then in the part of the DRC that is the subject of our ëtudy.

2.1.1. Geographic setting

The Batëkë plateau has already been described by several authors, BASAULA, (1989); BILOSO, (2008); BISIAUX et al, (2009); KASONGO et al, (2009); KASONGO, (2010). Of a ташёге дё^гак, it constitutes a vast structural ensemble that stretches from Gabon through the Rëpublique du Congo to the Rëpublique Dëmocratique du Congo.

It covers a total area of 35,164 km² of which 7,000 are in the DRC, bounded to the West by the Congo River, to the East at 17°00'longitude East, to the South by the 5°lat. South and to the North by the parallel 4°lat. South, (BASAULA, 1989). Its geographical aspect has hardly changed for centuries; it is an immense denuded space with some forest galleries lost in the middle of large grassy savannahs.

2.1.2. Geology

The Batëkë plateau forms the western extrem^ of the vast crëtacë a тшеёис age sedimentary basin that extends further east into the Rëpublique democratique du Congo. The oldest formations belong to the Stanley Pool group which rests directly on the Precambrian basement.

These are mostly soft, friable gres. The more recent strata are composed of sandy loam or quartz sand, resulting from in situ aeration and eolian sands. Two types of soil are distinguished in the region: hydromorphic soils and ferralitic soils on which the forests are established.

The soils are mostly sandy, very permeable and poor. According to KOUBOUANA et al, (2007), the geology of Plateau Teke is essentially two series: the Stanley Pool series generally found in Brazzaville and its northern surroundings and the Bateke sands series.

In the DRC, the Batëkë plateau is the name given to the northern extension of the Kwango plateau. In the east, the Kwango Plateau forms a structural surface with PHocёпе characteristics. This type of landscape is characterizedërisë by a monotonous regularity of plateau surface which is perfectly flat over large areas, (LADMIRANT, 1964; CAHEN and LEPERSONNE, 1948) cited by KASONGO, (2010).

The geomorphology is the same as that of the Bateke Plateau, interspersed with dry valleys and closed depressions of oval, lobed or sub-circular shape.

These depressions are rarely deeper than 20m. The large depressions are characterised by a muddy and wooded bottom; some are lined with a superficial formation of white sands (BASAULA, 1989).

The Plateau du Kwango massif, 600 to 700 m high, dominates the eastern part of the city and province of Kinshasa. Its portion located in the city is called Plateau des Bateke, the population density is very low. The chain of hills (350 to 675 m altitude) where the Ngaliema, Amba and Ngafula mountains are found, constitutes the common border with Central Kongo and forms the southern part of the city, up to the south-east, where the Bateke Plateau is found.

These hills, including the heights of Mbinza and Kimwenza, are thought to have arisen from the demantëlement of this Plateau. The Kinshasa plain follows the Congo River bed and is enclosed between the Congo River, the Bateke Plateau and the hills. It has an average width of only 5 to 7 km. This plain is situated between 300 and 320 m above sea level and has an area of about 100 km^2 .

This is where the largest portion of the population of the provincial city of Kinshasa is concentrated, (BILOSO, 2008). The soils of the Bateke Plateau were developed on the "Sables ocres" commonly referred to as the "Systëme de Kalahari", which includes the Neogëne series of the "Sables ocres" and the underlying Paleogëne series of the "Gres polymorphes", (LEPERSONNE, 1945; CAHEN and LEPERSONNE, 1956; De PLOEY et *al.*, 1968) in KASONGO, (2010). According to DE PLOEY (1965) quoted by BASAULA (1989), the Bateke Plateau region has a fairly typical geological profile: at the top, at about 690-700 m, the ochre sands of the Upper Kalahari are found.

This is followed by a transitional zone at around 610 m, which is formed by the Inftrieur Kalahari sands. Finally, at around 550 m, i.e. 150 m below the summit, we find the polymorphous sandstone and the soft mtsozoic shale.

2.1.3. Population
Often referred to as Ttkt (plural Battkt or singular Muttkt Teke according to the Africanist spelling), a Bantu people living in the western part of the Democratic Republic of Congo,

southern Congo and, to a lesser extent, south-eastern Gabon.

The Ttkts, as the tpithete indicates, are exclusively populated by an ethnic group subdivided into four tribes: the Ttkt, the Boma, the Nzikou and the Kukuya.

Their agricultural activities were mainly in the savannah, but the progressive impoverishment of savannah soils has led the Ttkts to practice forest agriculture (KOUBOUANA *et al.*, 2007).

The Ttkt people have no clan divisions, the main organisational principle is the chiefdom. The village is the most important social unit. The religion is the belief in the spirits of ancestors who never died and who are present with the living. As for the Higher Spirit, it inhabits the Lufimi River Falls for example.

It is the Nkwe Mbali that protects him. These spirits can act anywhere and in the service of the man who has an appropriate sanctuary. The Tktt society currently lives from the following activities: agriculture, charcoal burning, fishing, hunting, crafts and trade. On the Battkt plateau, cassava, maize, peanuts, potatoes and watermelon are grown in order of priority.

The Ttkt people are characterised by their tradition of hospitality, solidarity and peaceful determination.

2.1.4. Station in Ibi village

2.1.5. 1. Geographical situation

The Ibi-village site is located on the Batëkë plateau, which belongs administratively to the urban-rural Commune of Maluku, in the city-province of Kinshasa. It is located 140 km east of the centre of Kinshasa between 4°15' and 4°25' South latitude and 16°4' and 16°12' East longitude. The triangular-shaped Ibi-village station covers more than 20,000 ha.

It is bounded to the south by the Kinshasa-Kikwit road (the National No. 1), and to the west and east by the Duale and Lufimi rivers. It is reached 8 km from the main road and is easily accessible by a passable road. Historically, this area ë1я11 sparsely populated by ethnic Tëkë: about three inhabitants per кйотëйс square.

I . 'nitorite traditionnelle est exercëe par les chefs coutumiers dont le rôle, en droit moderne, n'est pas clairement précise en matiëre judiciaire et fon^re. Public order, hygiene, health, education and communications are the responsibility of the territorial authority (BISIAUX *et al.*, 2009).

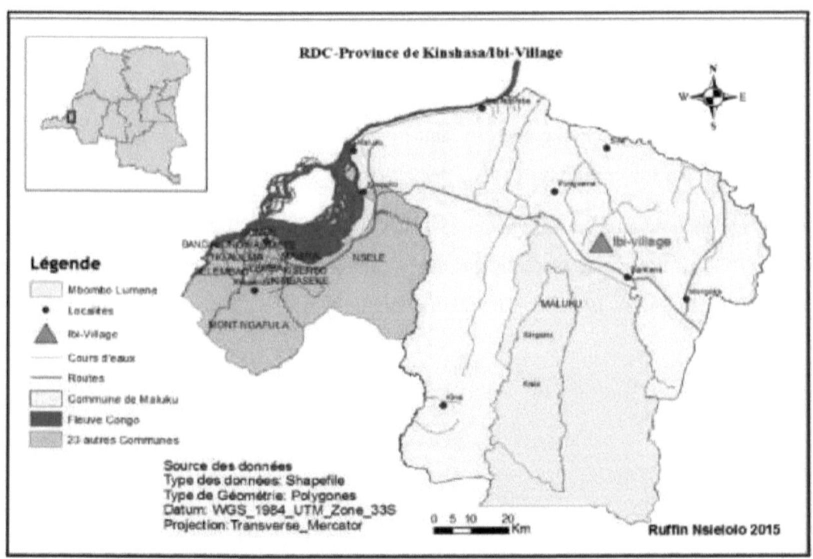

Figure 3: *Study environment (IBI). (Taken from NSIELOLO, 2015)*

2.1.4.2. Socio-economic activities

The population living in Ibi-village is composed of agroforestry workers living on the farm site with their families, plus the populations of villages, hamlets and farms within a radius of about 20 kilometres of the Ibi-village station.

In addition to locals, these villages, hamlets and farms are inhabited by migrants from Kinshasa or by the outflow of unemployed people from that city. The agroforestry initiative "Puits de Carbone Ibi-village" developed on the Bateke plateau has created about a hundred jobs. It can be an example of an intervention model to stabilise local populations or to fix migrants towards urban centres.

Apart from the temporary jobs offered by Ibi-village, the dominant activity in the region is itinerant bush farming, practised mainly in forest areas on the slopes of river valleys.

The exploitation of FFPs and NTFPs is well known on this plateau; the population also practises beekeeping, even if it is essentially artisanal. The cultivation methods linked to this agriculture imply the use of large areas to be burnt (1 to 4 hectares), which causes regular bush fires, damaging the natural forest regeneration.

Human presence has profoundly altered the environment over the past decade, although the effects of human disturbance are long lasting (KAYUMBA *et al.*, 2015).

2.1.4.3. Climatic factors

The climate is humid tropical, sudano-guinean of the Aw4 type according to the Koppen criteria, with an average annual precipitation of about 1500 mm from October to mid-May (HABARI *et al.*, 2010; VERMEULEN and LANATA, 2006). The dry season lasts four months, from mid-May to September; in January-February, and sometimes in mid-December, there is a period of lower rainfall.

The average annual temperature is about 25°C (Table 1) with absolute minima and maxima of 14 and 39°C in August and March respectively. But daily minimum temperatures in the dry season can be as low as 10-12°C. The relative humidity of the air is always quite high, 80% (BASAULA, 1989).

Early morning fog regularly stays on the Batëkë plateau sometimes until 9am. The Ibi site has only just installed its meteorological station, whose data do not yet cover a period of two years, so this site uses data from the Mampu and Mbankana stations located about 10 km away (Table 1).

Table 1: Average monthly weather data from the Mampu and Mbankana centre station for the period: 1997-2006

Station	Mampu		Mbankana	
Month	Prec. (mm/Month)	Temperature (°C)	Prec. (mm/Month)	Temperature (°C)
January	170,34	26,44	157,11	26,44
February	141,33	26,81	125,22	26,81
March	170,2	26,75	164,31	27,13
April	186,06	27,08	201,87	27,08
May	132,21	26,03	164,16	26,03
June	19,32	24,24	13,05	24,24
July	5,76	23,44	6,15	23,44
August	8,4	24,23	16,98	24,23
September	96,42	24,58	76,92	25,58
October	174,15	26,26	168,75	26,26
November	210,3	26,35	241,98	26,35
December	236,61	26,09	238,32	26,09
Maximum	**236,61**	**27,08**	**241,98**	**27,13**
Minimum	**5,8**	**23,4**	**6,2**	**23,4**
Average	**129,3**	**25,7**	**131,2**	**25,8**

Source: MANGONI, (2010)

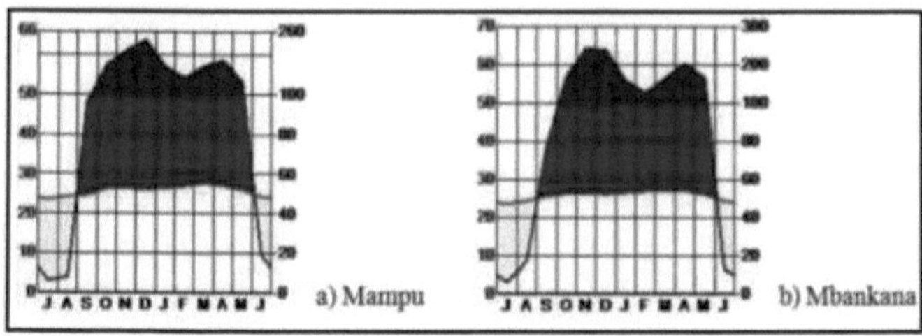

Figure 4: *Umbrothermal diagrams of Mampu (a) and Mbankana (b) stations*

The umbrothermal diagrams presented in Figure 2 show that the rainfall parameters of the Mbankana and Mampu stations have changed more or less regularly over the past 10 years. The average rainfall at Mbankana station is 2 mm higher than at Mampu.

Temperatures are slightly lower in Mampu, the difference between the mean temperatures is 0.1°C. A statistically insignificant difference. The climatic conditions are similar. The presence of Acacia forests in Mampu would justify this particular microclimate.

2.1.4.4. Hydrography

The Ibi-village concession is bordered on the West by the Duaie river, on the East by the Lufimi river and is crossed in its centre by the small Ibi river which gave its name (Ibi = spring) to the concession whose spring supplies the whole station with drinking water.

2.1.4.5. Vegetation

The Ibi station consists of grassy or shrubby savannahs that are rarely planted with trees. Two forest galleries form the eastern and western boundaries of the station on the banks of the Lufimi and Duale rivers, a third forest gallery occupies the slopes of the Ibi valley until it meets the Duale river to the north of the station.

Progressive dynamics are those characterised by the replacement of one vegetation by another in a series that tends towards the most evolved stage (climax), i.e. the vegetation that best corresponds to the ecological potential of the station in question; otherwise the evolution is regressive.

These modifications are reflected both on the physiognomic level, notably by the increase or reduction of strata, and on the floristic level by changes in species composition (MUHASHY *et al.*, 2011).

In the phytogeographic subdivision of tropical Africa, the Bateke Plateau region is located in the Guineo-Congolese region, Congolese domain, Bas-Congo sector, (WHITE, 1986 and BASAULA, 1989).

According to the same source, this vegetation has been the subject of several botanical and phytosociological studies, the main ones being those carried out by LEBRUN & GILBERT, (1954); MUAMBI (1968); COMPERE, (1970) and KALANDA, (1981). From these studies, we can derive a synthesis that allows us to define mainly three vëgëtal formations: the savannah, the forest and the aquatic vëgëtation.

a. Savannah vegetation

The savannahs include grassy savannahs with *Digitaria longiflora* and *Hyparrhenia diplandra*, as well as shrubby savannahs with *Hymenocardia acida*, *Crossopteryx febrifuga*, *Annona senegalensis* and *Bridelia feruginea* covering vast ëtends.

b.   Forest vegetation

The valley bottoms are occupied by gallery forests along the Duale and Lufimi rivers. These forests are characterised by species such as *Millettia laurentii* (an over-exploited species in these gallery forests that remains only as bushy stumps), *Pentachletra eetveldeana*, *Leptactina leopoldii*, *Hymenocardia ulmo'ides*, *Oncoba welwitschii*, *Rauwolfia mannii*, *Macaranga spinosa, M. monandra*, etc.

Despite the efforts of NOVACEL and GI-Agro to conserve habitats and overall biodiversity, these forests are quite often destroyed by fire and clearing. When these threats cease, the reconstitution of these habitats remains possible in many places, (MUHASHY, 2011).

2.1.4.6.   Wildlife

The potential fauna to be found in Ibi- village can be considered to be that of the neighbouring Bombo Lumene Reserve. These are mainly Savannah buffaloes (*Syncerus caffer*), Harnessed guib (*Tragelaphus scriptus*), Sitatunga (*Tragelaphus spekei*), (MUHASHY, 2011).

Potamocheres, Jackals, Monkeys, birds. This fauna is subject to intensive poaching, even in the reserve (NOVACEL, 2009). We also note the presence of the mole rat (*Cryptomys mechowi*), a rodent that lives in the ground; known under the local name of Pombofuko in the province of Haut-Katanga, Tshibulebule in the province of Kasai Oriental (Mbuji-Mayi) and Ikwe in the Bateke plateau, (KISASA et *al.*, 2005).

Antelopes also often frequent the contact zone between savannah and forest (MUHASHY, 2011).

The birdlife is very rich and varied, with partridges being the most common birds encountered along the route. Bush fires frequently lit in the savannah and hunting are the main cause of animal migration.

Once very present in the region, the large fauna (buffalos, lions, etc.) has been decimated by hunting. Even the small herbivores are in danger of extinction due to the incessant hunting.

2.1.5. Presentation of the Ibi-village arboretums

Here we briefly present the arboretums containing our studied species.

2.1.5.1. South Arboretum (near the enclosure)

Installed in 2015 on an area of 1500 m^2 or 0,15 ha, this arboretum has in its whole 5 species of edible caterpillar trees divided into 5 families and counts 13 individuals.

2.1.5.2. The North Arboretum (creiir)

It has an area of 2100 m2 or 0.21 ha and includes 5 species of caterpillar trees divided into 2 families with 9 individuals.

2.1.5.3. Arboretum 3 North

This arboretum was established in 2015, it is 50 m long and 40 m wide, i.e. 2000 m^2 (0.20 ha) and contains 11 species of caterpillar trees divided into 8 families with 29 individuals.

2.1.5.4. South Arboretum (extension)

The southern extension arboretum was established in 2017 on an area of 10,000 m2 or 1 ha. It includes 10 species of caterpillar trees divided into 8 families and has a total of 24 individuals.

2.1.5.5. Arboretum caterpillar wood

It was installed in December 2017 on a 0.8 ha plot 30 m south-east of the Residence: 109 individuals of caterpillar trees belonging to 10 different species and grouped into 8 families were planted.

2.2. Material

The biological matëriel ий^ë in the realisation of this study consists essentially of herbarium samples of 35 species of caterpillar plants planted in the SAF of lbi-village as well as some caterpillars collected in this study environment identified by BELESI and LEJOLY according to Latham's book (2008).

For the realisation of this study we used the following tools:

- the machete: which allowed us to open up passages.
- a shaker: for taking herbarium samples.
- a tape measure: to measure the circumference of trees.
- 1.30 cm rod: to measure the height of trees at 1.30 m from the ground.
- a techno camon 12 mobile phone with on-board camera: to capture photos
- a computer: for the writing of this work
- a 10 m bamboo: to take the height of the trees
- a research notebook: to record our fieldwork

-       a pen: to write down our fieldwork
-       newspaper, cardboard and a wooden frame: to make our herbarium.

2.3.   Methods

Our working method is based on observation. These observations included the identification of caterpillar plants, biological type, diaspore type, leaf type, morphological type and the collection of some of the said caterpillar host plants, followed by a small survey among the inhabitants of ibi-village, the processing and analysis of these data to obtain results.

To do this we proceeded as follows:

2.3.1.   Direct observation in the field

2.3.1.1.   Site survey

With the help of the research certificate obtained from the Dean of the Faculty of Sciences, observations and visits were organised from 26 July to 12 November 2020 in the Bateke plateau, more precisely in Ibi-village in the urban-rural commune of Maluku, in the city of Kinshasa. These visits were carried out with the aim of reconnoitring the study area in order to collect information related to the theme being addressed. These are the aspects related to climatic, ëdaphical factors, etc.

2.3.1.2.   Floristic inventory

The floristic inventory was done on all edible caterpillar plants planted in the agroforestry system of Ibi-village.

a. Taking diameters and heights

The tree diameter was obtained by measuring the circumference of each tree with a tape measure and dividing it by П or 3.14 to 1.30 m from the ground above the stilt roots, buttresses and below the first branching for plants over 3 m tall and at the collar for trees less than 3 m (LOKOMBE, 2007).

This height was achieved by using a 1.30 m high stick. The height of the trees was measured with a 10 m graduated bamboo.

2.3.1.3.   Species identification

The botanical samples thus collected were submitted for identification to the laboratory of Systemics, Biodiversity, Nature Conservation and Endogenous Knowledge using the Flora of Africa, Cameroon, Rwanda - Urundi (vol. I to X and some fascicles). These samples have been classified according to the APG II, III, and IV system, verified and confirmed by the director of this thesis.

2.3.1.4. Autoecological characteristics

a. Phytogeographical distribution (P.D.)

The phytogeographical groups of the inventoried species allow us to determine the phytogeographical position of our work. Referring to the phytogdographic subdivisions of Central Africa, as proposed by WHITE (1979, 1986), we have retained the following categories:

- Guinean-Congolese (G.C.): observed throughout the Guinean-Congolese region or multi-domain species; e.g. *Martretia quadricornis* Beille.
- pantropicales (Pan): species found in Africa, America, tropical Asia and Australia (intertropical region). Ex : *Ceiba pentandra* (L.) Gaertn

- bas-guiito-congolais (BGC): present in the bas guiittn and Congolese sub-centres; distributed from the Хьдёпа in the Rёpublique Dёmocratique du Congo, e.g. *Staudtia kamerunensis* Warb. var. gabonensis (Warb.) Fouilloy.
- Afro-tropical (A.T): found in several phytochories in continental tropical Africa. Ex: *Pseudospondias microcarpa* (A. Rich) Engl.
- Guiito-Eongolian Centre (GEC): is a species with a tributary area that does not reach the upper Guiiten domain.

b. Biological type (T.B)

The examination of biological types allows us to determine the adaptive strategies, as well as the physiognomy of the vёgёtation. In our work, we have adopted the biological types defined according to the classification of RAUNKIAER (1934) modified by LEBRUN (1947), namely :

- mёsophanёrophyte (MsPh): trees whose organs are located between 10 and 30 m above the ground
- mёgaphanёrophyte (MgPh): trees whose organs are located above 30 m from the ground.

c. Type of diaspore (T.D)

Diaspore types provide information on the mode of dissёmination of species. 4 categories of diaspore defined by DANSEREAU & LEMS (1957) have been retained for our work, they are :

- ballochores (Ballo): diaspores expelled by the plant itself,
- bogonochores (Pugno) : Diaspores with feathery or silky appendages, egret hairs
- barochores (Baro): dry or fleshy, heavy diaspores
- sarcochores (sarco): totally or partially fleshy diaspores.

d.   Leaf type (T.F)

The types of leaf sizes were ëtë inspired by RAUNKIAER (1934), taken up by many authors, including MALAISSE (1976), LUBINI (1997,2001), MASSENS (1997), NSHIMBA (2008), BELESI (2009), HABARI (2009) and RUELLE & MALAISE (2016).

These are the following types:

- mësophylls: leaf area between 20 and 200 cm^2 (2 dm^2),
- leptophylls: leaf area less than 0.2 cm2,
- mëgaphylls: leaf area greater than 0.16 m2
- macrophylls: 2 to 200 dm2
- nanophylls: 0.2 to 2 cm2
- microphylls: 2 to 20 cm2

e.   Morphological type (M.T.)

The morphological spectra identified in the phytosociological surveys have ëtë dëterminatedës in the field and their vërification has ëtë rëalisëe using the computerised catalogue of LEJOLY et *al.* (1988).

The morphological types selected for this work are :

- **tree (A)**: a species with a woody stem, of large size with the presence of axillary buds that rise several metres.

- **shrub (Arb)**: a species with a woody stem, small in size, usually with the absence of axillary buds rising a few metres and branching at their base.

2.3.2.   Survey of the inhabitants of lbi-village

Our surveys were conducted among the inhabitants of Ibi, some of them were interviewed in Musiabikui and others in the workers' hostel in Ibi-village.

To do this, a survey protocol (see annex) was ëtë ëlaborë and submitted to 10 people living in Ibi.

The criteria for sëlection of enquëtës are as follows:

- to have lived at least for a long time in d'ibi
- to know Ibi and its surroundings perfectly
- know the vernacular names of caterpillars and their host plants;
- know the përiodicitë of the caterpillar harvest.

2.3.3. Analysis and interpretation of data

For the processing and interpretation of the data, we used Microsoft Word, Excel and some statistical analysis.

For the statistical analyses, we analyzed the following parameters according to BILOSO (2021):

• **Percentage (%):** this is calculated by dividing the frequency (f) of each individual by the total number (N) of all individuals multiplied by 100.

$$\% = \tfrac{\wedge}{N} * 100$$

• **the range of variation (E.V):** it is calculated by taking the maximum value of the classes minus the minimum value of the classes and adding one unit.

E. V= (Vmax-Vmin) +1

• **number of classes (k):** this is calculated according to the Sturges formula.

k=1+3.3logN

But for our case, we decided to use k=3 for the simple reason that for our purposes, we decided to classify the data into three classes (the worst performing, the average and the best performing in height and diameter) and our classes were ranked in ascending order of size (from smallest to largest).

• **class interval (i):** it is found by dividing the range of variation (E.V) by the number of classes (k).

For the annual average of our species, we took the average growth of each species in the arboretum in height and diameter, divided by the number of years since the species were planted and finally we took the annual growth of each species that recurs in each of the arboretums, we summed each annual growth divided by the total number of individuals of that species.

However, we note that the raw data for the graphs are presented in the annexed tables.

Photo 1: During our data collection (photo KIDUSHI, 2020).

CHAPTER 3

RESULTS AND DISCUSSION

In this chapter, we present our results in the form of tables and figures, specifying the families, species, percentages of autoëcotogical characteristics of each species, the performance of the species studied and the potentialities of each species. We will discuss the results with other authors in relation to previous studies conducted.

3.1. Results

3.1.1. Ethnobotanical aspect

3.1.1.1. Edible caterpillar plants

The edible caterpillar plants planted in Ibi-village are listed in Table 2.

Table 2: Diversity of caterpillar host plants domesticated in Ibi-village.

ORDER	FAMILY	SPECIES
Asterales	Asteraceae	*Vernonia brazzavilensis* Schreb.
Fabales	Fabaceae	*Afzelia bipendensis* Harms. *Afzelia bella* Harms. *Albizia adianthifolia* Benth. *Berliniagiorgii* Sol. Hook.f. *Millettia drastica* Wight & Arn. *Millettia laurentii* De wild. *Millettia versicolor* Wight & Arn. *Pentaclethra eetveldeana* De wild. T. Durand *Pentaclethra macrophylla* Benth.
Gentianales	Apocynaceae	*Funtumia elastica* (preuss) stapf. *Picralina nitida* (stapf) T. Durand & H. Durand.
	Loganiaceae	*Anthocleista schweinfurthii* Gilg.
Malpighiales	Rubiaceae	*Sarcocephalus latifolius* (Sm) E.A. Bruce. *Nauclea diderrichii* Merr.
ORDER	**FAMILY**	**SPECIES**
Malpighiales	Euphorbiaceae	*Aleurites mollucana* (L.) wild. *Bridelia feruginea* Benth. *Hymenocardia ulmoides* Wall & Lindl. *Macaranga spinosa* Thouars. *Ricinodendron heudelotii* Mull. Arg. *Uapaca guineensis* Mull. Arg.
	Flacourtiaceae	*Oncoba welwitschii* Forssk.
	Irvingiaceae	*Irvingia gabonensis* Baill.
	Clusiaceae	*Symphonia globulifera* L.f.
Malvales	Malvaceae	*Ceiba pentandra* L.
Myrtales	Myrtaceae	*Syzygium guineensis* (Wild).
Sapindales	Burseraceae	*Canarium schweinfurthii* Engl. *Dacryodes edulis* (G. Don) H.J. Lam.

	Anacardiaceae	*Lannea antiscorbutica* A. Rich.
	Meliaceae	*Entandrophragma angolense* (Welw.) C.DC.
		Entandrophragma candolleii Harms.
Hives	Moraceae	*Milicia excelsa* (Welw.) C.C Berg.
		Cercopiapeltata L.
		Trilepsium madagascariensis DC.
Rosales	Rhamnaceae	*Maesopsis eminii* Engl.

The analysis of the above table shows that 35 species of edible caterpillar plants were rëpertoriëed and belonging to 7 orders and in 16 families.

The orders *Fabales* and *Malpighiales* are best represented with 9 species each, followed by *Sapindales* with 5 species and *Gentianales* with 3 species and the others are poorly represented.

From the family point of view, the *Fabaceae* and *Euphorbiaceae* families are dominant, with 9 and 5 species respectively, and the others are weakly represented.

3.1.1.2.   Density of edible caterpillar host plants planted in SAF in Ibi-village.

The quantitative analysis of the species inventoried is shown in Table 3 below.

Table 3: Quantitative analysis of host plants of domesticated edible caterpillars in FAS in Ibi-village.

Legend: f: Frequency, **%:** Percentage

N°	SPECIES	FAMILY	f	%
1	*Afzelia bipendensis* Harms.	Fabaceae	4	2
2	*Afzelia bella* Harms.	Fabaceae	8	4
3	*Albizia adianthifolia* Benth.	Fabaceae	2	1
4	*Aleurites mollucana* (L) wild.	Euphorbiaceae	9	5
5	*Anthocleista schweinfurthii* Gild.	Loganiaceae	2	1
6	*Berlinia giorgi* Sol.	Fabaceae	11	5,7
7	*Bridelia feruginea* Benth.	Euphorbiaceae	1	0,5
8	*Canarium schweinfurthii* Engl.	Burseraceae	8	4
9	*Ceiba pentandra* L.	Malvaceae	3	1,5
10	*Cercopia peltata* L.	Moraceae	1	0,5
11	*Dacryodes edulis* (G. Don) H.J Lam.	Burseraceae	4	2
12	*Entandrophragma angolense* (welw.) C. DC.	Meliaceae	3	1,5
13	*Entandrophragma candolleii* Harms.	Meliaceae	2	1
14	*Funtumia elastica* (Preuss) stapf.	Apocynaceae	12	6,2
15	*Hymenocardia ulmo'ides* Wall & Lindl.	Euphorbiaceae	1	0,5
16	*Irvingia gabonensis* Baill.	Irvingiaceae	6	3,1
17	*Lannea antiscorbutica* A. Rich.	Anacardiaceae	1	0,5
18	*Macaranga spinosa* Thouars.	Euphorbiaceae	2	1
19	*Maesopsis eminii* Engl.	Rhamnaceae	12	6,2
20	*Milicia excelsa* (Welw) C.C Berg.	Moraceae	16	8
21	*Millettia drastica* Wight & Arn.	Fabaceae	1	0,5
22	*Millettia laurentii* De wild.	Fabaceae	1	0,5
23	*Millettia versicolor* Wight & Arn.	Fabaceae	2	1
24	*Nauclea diderrichii* Merr.	Rubiaceae	4	2
25	*Oncoba welwilschii* Forssk.	Flacourtiaceae	1	0,5
26	*Pentaclethra eetveldeana* De wild. T. Durand.	Fabaceae	39	20,1
27	*Pentachletra macrophylla* Benth.	Fabaceae	9	5
28	*Picralina nitida* (stapf) T. Durand & H. Durand.	Apocynaceae	1	0,5
29	*Ricinodendron heudelotii* Mull. Arg.	Euphorbiaceae	9	5
30	*Sarcocephalus latifolius* (Sm) E. A. Bruce.	Rubiaceae	1	0,5
31	*Symphonia globulifera* L. f.	Clusiaceae	1	0,5
N°	SPECIES	FAMILY	f	%
32	*Syzygium guineensis* (wild).	Myrtaceae	1	0,5
33	*Trilepsium madagascariensis* DC.	Moraceae	7	3,6
34	*Uapaca guineensis* Mull. Arg.	Euphorbiaceae	5	2,6
35	*Vernonia brazzavilensis* Schreb.	Asteraceae	4	2
	Total		**194**	**100**

Table 3 shows the dominance of *Pentaclethra eetveldiana* with 39 individuals or 20.1% followed by *Milicia excelsa* with 16 individuals or 8% then *Maesopsis eminii* and *Funtumia elastica* follow with 12 individuals each or 6.2% and the others are weakly represëntëes.

3.1.1.3. Autoecological characteristics of edible caterpillar plants

Ibi

The autoëcological characteristics of edible caterpillar plants planted in ibi-village SAF are presented in this section.

a.   Phytogeographical distribution

The different geographical distributions of the caterpillar species planted in Ibi-village are shown in figure 5.

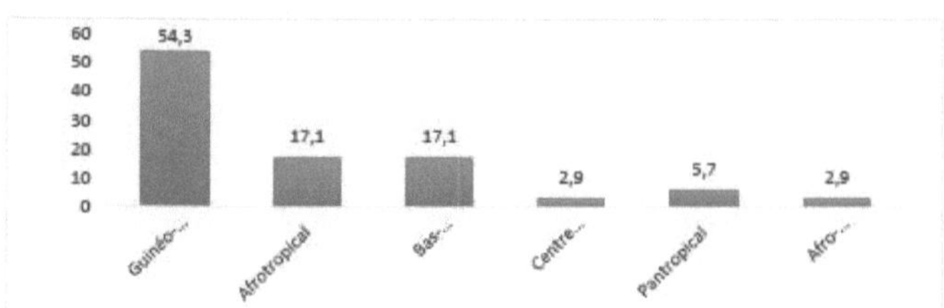

Figure 5: Phytogeographic spectra of the caterpillar species studied

In the analysis of figure 5, we note the dominance of species belonging to the Guinean-Congolese regional centre of endemism at 54.4%, i.e. 19 species, followed by the Afrotropical and Lower Guinean-Congolese species (17,1% each), the Pantropical species (5.7%) and the Central Guinean and Afro-American species are poorly represented with 2.9% each (i.e. 1/35 species each). Therefore, the species that characterise this plantation are those of the Guinean-Congolese regional centre of endëmism. Biological types

The analysis of the biological types of species is shown in the figure below below.

Figure 6: Biological types of caterpillar host plants in SAF d'Ibi-village.

Examination of Mesophanerophytes (85.7%) Figure 6 shows the dominance of the species i.e. 30/35 species) and the other species are Megaphanerophytes, i.e. 14.3%.

b.   Types of diaspora

Figure 7 explains how the species belong to the different diaspora types

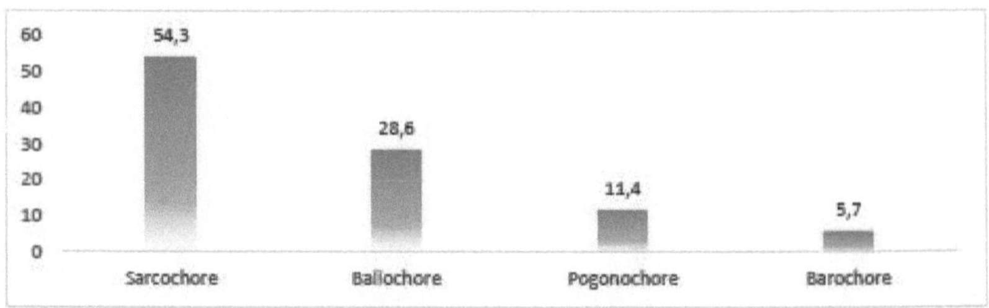

Figure 7: types of caterpillar tree diaspores planted in the SAF of lbi-village.

From this figure 7, it appears that sarcochory is the most dominant mode of dispersal of the diaspore with 53% of the species studied followed by ballochory with 31%. Pogonochory and barochory are poorly represented with 10% and 6%.

c.   Leaf type

The different types of leaf size are shown in figure 8.

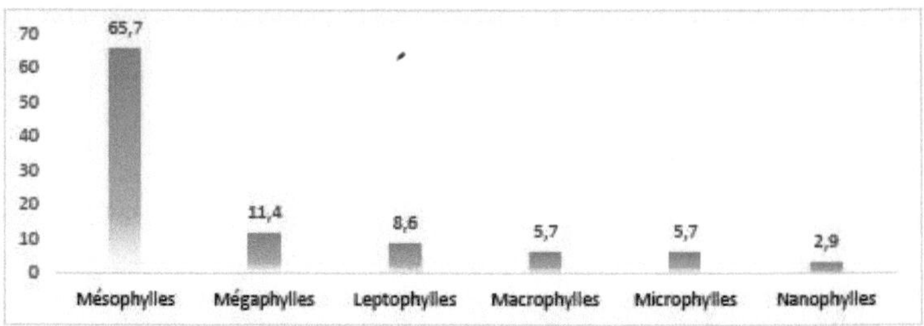

Figure 8: Leaf types of caterpillar trees planted in SAF d'lbi-village.

From figure 8 it can be seen that Mesophylls dominate at 65.7% followed by Megaphylls (11.4%), Leptophylls with 8.6%. Macrophylls and Microphylls follow with 5.7% and Nanophylls 2.9%.

d.   Morphological types

Two morphological types are equally represented (Fig.9)

Figure 9: Morphological types of species planted in SAF d'Ibi-village.

Our analysis shows that the proportion of trees is very dominant with 91.4% while shrubs are less represented (8.6%).

3.1.2. Performance of edible caterpillar plant species planted in SAF of Ibi-village.

In this section we classify the species into three categories: the worst performing, the average performing and the best performing.

3.1.2.1. Performance in Height

On this point, we classify the species according to the annual performance of each. In m/year for height, in cm/year for diameter.

a. **Least performing species** (0.2 to 1.3 m per year).

Table 4 shows the heights of the worst performing caterpillar species.

Table 4: Average growth of the worst performing species in height over one year

N°	Species	Height (m/year)
1	*Afzelia bipendensis* Harms.	0,2
2	*Symphonia globulifera* L.f.	0,2
3	*Anthocleista schweinfurthii* Gild.	0,3
4	*Berlinia giorgii* Sol.	0,3
5	*Millettia versicolor* Wight & Arn.	0,3
6	*Oncoba welwitschii* Forssk.	0,3
7	*Sarcocephalus latifolius* (Sm.) E. A. Bruce.	0,3
8	*Macaranga monandra* Thouars.	0,4
9	*Vernonia brazzavilensis* Schreb.	0,4
10	*Afzelia bella* Harms.	0,5
11	*Entandrophragma angolense* (Welw.) C. DC.	0,5
12	*Nauclea diderrichii* Merr.	0,6
13	*Trilepsium madagascariensis* DC.	0,6
14	*Irvingia gabonensis* Bail.	0,7
15	*Pentaclethra eetveldeana* De wild. T. Durand.	0,7
16	*Picralina nitida* (Stapf) T. Durand. & H. Durand.	0,7

17	*Uapaca guineensis* Mull Arg.	0,7
18	*Aleurites mollucana* (L.) Wild.	0,8
19	*Entandrophragma candolleii* Harms.	0,8
20	*Milicia excelsa* (welw.) C.C. Berg.	0,8
21	*Dacryodes edulis* (G. Don.) H.J. Lam.	0,9
22	*Funtumia elastica* (Preuss.) Stapf.	0,9
23	*Pentaclethra macrophylla* Benth.	0,9
24	*Ricinodendron heudelotii* Mull Arg.	0,9
25	*Syzygium guineensis* (Wild.).	0,9
26	*Hymenocardia ulmo'ides* Wall & Lindl.	1
27	*Millettia drastica* Wight & Arn.	1
28	*Bridelia feruginea* Benth.	1,2
29	*Cecropiapeltata* L.	1,3

Analysis of Table 4 shows that the class of poorest performing species comprises 29 species, which represents 83% of the ëtudiëe flora. Among the species with poor growth performance, we can note *Afzelia bipendensis*, *Symphonia globulifera*, etc.

b. **Species with average performance** (1.4 to 2.5 m/year).

Table 5 below explains the variation in height of species with average performance of 1.4 to 2.5 m/year.

Table 5: average of caterpillar species with average performance during one year.

N°	Species	Height (m/year)
1	*Ceiba pentandra* L.	1,5
2	*Canarium schweinfurthii* Engl.	1,7
3	*Maesopsis eminii* Engl.	1,8
4	*Millettia drastica* Wight & Am.	2
5	*Albizia adianthifolia* Benth.	2,2

The analysis of Table 5 shows that the class of species with medium height performance includes 5 species, representing 14.2% of the flora studied. These include *Ceiba pentandra*, *Canarium schweinfurthii*, etc.

c. **Best performing species** (2.6 to 3.7 m/year).

Our study shows that in the class of the best performing species, we find only one species: *Lannea antiscorbutica* with an annual growth of 2.7m/year and it represents only 2.8% of the studied flora.

3.1.2.2. Diameter performance of caterpillar plants planted in SAF d'lbi- village. a. Lists of the worst performing species (0.4 to 1.9 cm/year)

Table 6 shows the diameters of each species studied.

Table 6: Annual average of the worst performing species

N°	Species	Diameter (Cm/year)
1	*Afzelia bipendensis* Harms.	0,4
2	*Oncoba welwitschii* Forssk.	0,5
3	*Afzelia bella* Harms.	0,6
4	*Funtumia elastica* (Preuss.) Stapf.	0,9
N°	Species	Diameter (Cm/year)
5	*Hymenocardia ulmo'ides* Wall & Lindl.	0,9
6	*Irvingia gabonensis* Baill.	0,9
7	*Picralina nitida* (Stapf.) T. Durand & H. Durand.	0,9
8	*Trilepsium madagascariensis* DC.	0,9
9	*Anthocleista schweinfurthii* Gilg.	1
10	*Macaranga monandra* Thouars.	1
11	*Uapaca guineensis* Mull. Arg.	1
12	*Vernonia brazzavilensis* Schreb.	1
13	*Entandrophragma candolleii* Harms.	1,1
14	*Sarcocephalus latifolius* (Sm.) E. A. Bruce.	1,1
15	*Symphonia globulifera* L.f.	1,1
16	*Pentaclethra eetveldeana* De Wild. T. Durand.	1,2
17	*Berlinia giorgi* Sol.	1,3
18	*Millettia laurentii* De Wild.	1,3
19	*Millettia versicolor* Wight & Arn.	1,4
20	*Aleurites mollucana* (L.) Wild.	1,6
21	*Dacryodes edulis* (G. Don.) H.J. Lam.	1,6
22	*Pentaclethra macrophylla* Benth.	1,6
23	*Nauclea diderrichii* Merr.	1,7
24	*Syzygium guineensis* (Wild.)	1,8
25	*Milicia excelsa* (Welw.) C.C. Berg.	1,9

Table 6 shows that the class of species with a low performance in diameter numbered 25 and represented 71.4%. These include *Afzelia bipendensis*, *Oncoba welwitschii*, *Afzelia bella*, etc.

b. Species with average performance (2.0 to 3.5 cm/year).

Table 7 shows the diameters of the medium-performing tracked species.

Table 7: List of medium performing species.

N°	Species	Diam. (Cm/year)
1	*Canarium schweinfurthii* Engl.	2,1
2	*Cercopiapeltata* L.	2,1
3	*Ricinodendron heudelotii* Mull. Arg.	2,1
4	*Entandrophragma angolense* (Welw.) C. DC.	2,8
5	*Lannea antiscorbutica* A. Rich.	2,8
6	*Millettia drastica* Wight & Arn.	3,1

The analysis of Table 7 shows that the class of species with medium performance in diameter includes 6 species or 17.1%. These include *Canarium schweinfurthii*, *Cercopia peltata*, *Ricinodendron heudelotii,* etc.

c. High performance species (3.6 to 5.1 cm/year).

Table 8 shows the diameters of the tracked species with high diameter performance.

Table 8: List of high performing species with one year of age.

N°	Species	Diam. (Cm/year)
1	*Bridelia feruginea* Benth.	3,6
2	*Maesopsis eminii* Engl.	3,6
3	*Albizia adiantifolia* Benth.	3,7
4	*Ceiba pentandra* L.	4,2

The analysis of Table 8 shows that the species with a high performance in diameter are 4 in number, i.e. 11.4% of the ëtudiëe flora. Among these species with the best performance in diameter, we can note *Bridelia feruginea, Maesopsis eminii*, etc.

3.1.3. Ethnozoological aspect.

3.1.3.1.Distribution of the surveys according to the survey site

The distribution of investigëtës by survey site is recorded in Table 9.

Table 9: Distribution of surveys by survey site

N°	Survey site	Frequency	%
1	Musiabikui	5	50
2	Workers' hostel	5	50
	Total	**10**	**100**

The analysis of Table 9 shows that our survey sample is 10 and is divided into two groups. The first group of 5 in Musiabikui and the second group of 5 in the workers' hostel near the ibi spring.

It appears from these 100% surveys that the main caterpillar collected in large quantities in ibi-village is *Imbrasia ertli* (Mimbimbi). But also two other edible caterpillars *Cymothoe caemi* and *Antheua insignata* are often collected in ibi-village and its surroundings and that the harvest extends from February to December.

3.I.3.2. Caterpillar species observed in Ibi and its surroundings

The taxonomic identification of the caterpillars is illustrated in the table below.

Table 10: Systematic group of edible caterpillars identified.

Branch	Class	Order	Family	Species
Arthropods	Insects	Lepidopterans	Notodontidae	*Imbrasia ertli*
				Antheua insignata Gaede
			Nymphalidae	*Cymothoe caemis* Drury

Overall, 3 species of edible caterpillars were recorded in our study area. They are grouped in 2 families: the best represented family is the Notodontidae with two species or 66.6% and the others are poorly represented. This is best illustrated in figure 10.

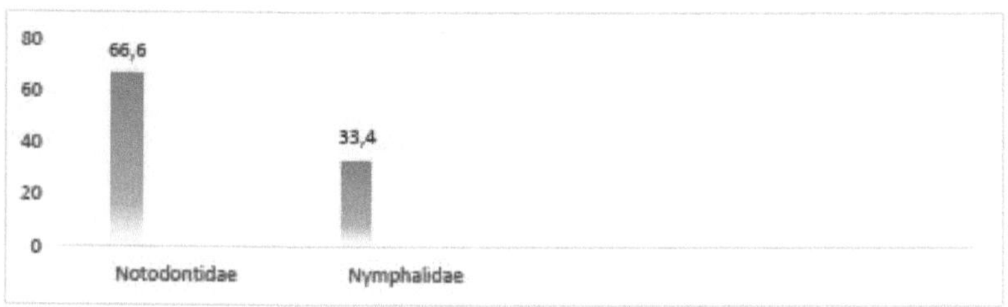

Figure 10: different families of caterpillars identified

3.I.3.2. Some caterpillars and their host plants

The caterpillars and their host plants are listed in Table 11. Each caterpillar species has, in addition to the scientific name, a vernacular name for the region under study.

Table 11: caterpillars and host plants

Edible caterpillars		Food plants	
Scientific names	Vernacular names	Scientific names	Vernacular names
Antheua insignata	Nsangu Mindanda (Kikongo Kwilu)	*Hymenocardia ulmoides*	Munsanga (Kikongo)
		Millettia drastica	Ngilu (Kikongo)
Imbrasia ertli	Mimbimbi (Kikongo)	*Pentaclethra eedveldeana*	Miigway (yansi)
Cymothoe caemi	Mibamba, mita (Kikongo)	*Oncoba welwitschii*	Mubamba (Kikongo)

Analysis of table 11 shows that most of the caterpillars collected in Ibi-village and its surroundings have a polyphagous and monophagous diet, *Antheua insignata* for example nests on *Hymenocardia ulmoides* and *Millettia eedveldeana* unlike Cymothoe caemi which

45

has a monophagous diet; it only feeds on the leaves of *Oncoba welwitschii*

3.I.3.3. Some photos of the caterpillars studied

1.   Cymothoe caemi

This common species is not much appreciated because of its bitter taste. In the region of Kinshasa, Central Kongo and Kwilu, it nests on *Oncoba welwitschii* (Mibamba) and can be collected in large quantities in July (LATHAM, 2008).

Photo 2: *Photo of Cymothoe caemiprise a Ibi (Musiabikui) 2020 (Photo Kidushi 2020)*

2.   Antheua insignata

Antheua insignata (Photo 2) is a very important edible caterpillar that is collected in December and January. It can be stored whole (fresh or dried) for consumption. Its food plant is Munsanga or Munsaya (*Hymenocardia ulmoides*) and Ngilu (*Millettia drastica*). Often harvested in December (LATHAM, 2008).

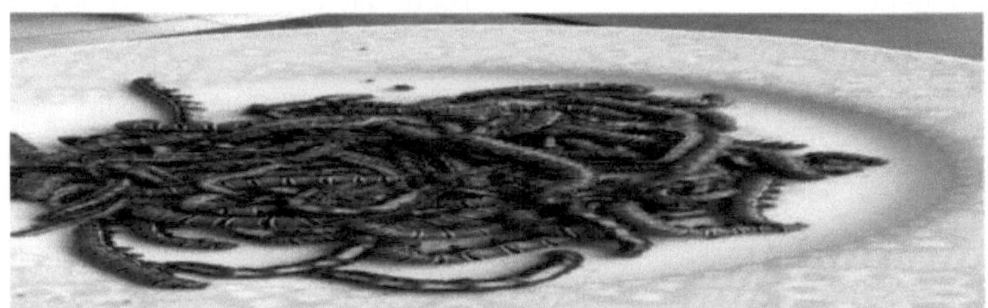

Photo 3: Antheua insignata in a container in Ibi-village (Musiabikui) (Photo Kidushi 2020)

3.   Imbrasia ertli

While most caterpillar species remain in one place as long as they are not destroyed by overharvesting, fire or loss of the plant on which they feed, *Imbrasia ertli* appears in large numbers in completely different places from year to year.

It first appears on the Kiseka (Pentaclethra eedveldeana) but will soon feed on most of the plants it finds around it.

If the caterpillars are on plants within arm's reach, they can be called by shouting "Hey-hey or bii-bii" in a low voice. This will cause them to move and thus be identified by the collector. On the other hand, if they are on high branches, attempts are sometimes made to smoke them in order to make them fall. If the host species is not very high, one climbs on it to shake the branches to let them fall to the ground in order to pick them up.

The caterpillars are collected from October to December in Central Kongo and from May to June in the Bateke plateau, only in forests that have been left to grow for some time and then in the shrubby grassland formation and on acacias. In the past, traditionally, after the Mimbimbi harvest, the village chief gave the order to cut down the forest in order to plant crops. (LATHAM, 2008).

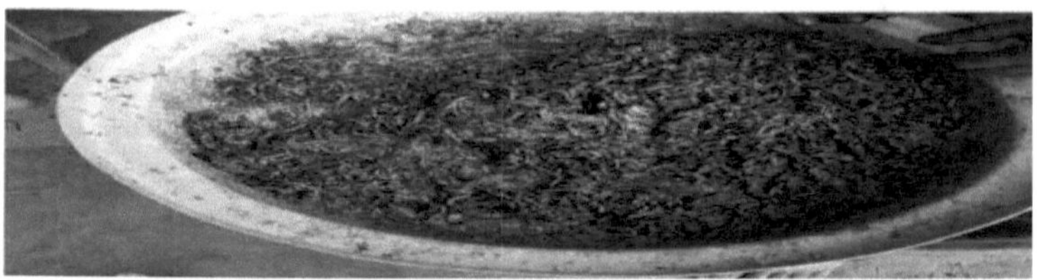

Photo 4: Imbrasia ertli caterpillar photographed during the June harvest in Ibi-village at Musiabikui.
Ibi-village at Musiabikui (Photo Kidushi 2020)

3.1.4. Other potentialities of caterpillar plants studied.

In this paragraph, we present the potential alternative uses of caterpillar plants planted in the lbi-village FAS and the potential caterpillars that may feed on them in the future.

3.1.4.1.   Different uses of caterpillar plants planted in SAF of Ibi-village.

The different uses of caterpillar plants planted in the ibi-village FAS are shown in figure 11.

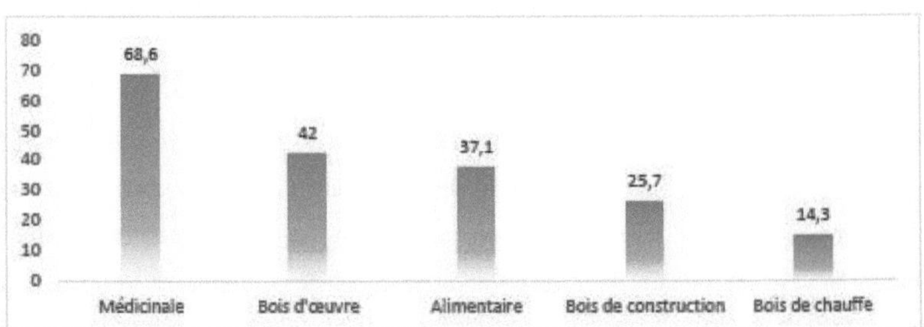

Figure 11: Different uses of caterpillar plants planted in SAF in Ibi-village.

Analysis of Figure 11 shows the prevalence of species used for medicinal purposes with 24 species out of 35 or 68.6%, followed by species used for timber with 15 species or 42.9% and food use with 13 species out of 35 or 37.1%. The construction wood group with 9 species out of 35 (25.7%) and fuelwood with 5 species out of 35 (14.3%) are the last categories.

These species have multiple uses. A single species can have several uses and play several roles. For food uses, these species give edible fruits such as *Dacryodes edulis, Sarcocephalus latifolius, Symphonia globulifera*, etc.

Others are medicinal thanks to the virtues of their leaves, fruits, essential oil, bark and roots. For the species used for construction, the sticks are used to build houses in villages. The timber is used to make boards for formwork and other uses in carpentry and construction. As for the plants used as firewood, the dried branches are used to light fires in rural and even urban areas.

3.1.4.2. Possible future caterpillars of species planted in Ibi-village.

On this point, we will build on previous research to try to identify caterpillars that feed on caterpillar plants planted in Ibi village.

Table 13: Possible caterpillars that may nest on host species of caterpillars

N°	Caterpillar species Potential	Vernacular names	Food plants
1	*Imbrasia petiveri*	Mimbermben (Y ansi)	*Milicia excelsa* (Welw.) C.C. Berg.
			Ricinodendron heudelotii Mull. Arg.
2	*Bunaea alcinoe*	Lukowo (Kikongo Central Kongo)	*Sarcocephalus latifolius* (Sm.) E.A. Bruce.
			Dacryodes edulis (G. Don.) H.J. Lam.
			Uapaca guineensis Mull. Arg.
3	*Imbrasia ertli*	Mindanda (Kikongo Kwilu) Endan/Essiir (Yansi Kwilu)	*Millettia laurentii* De wild.
			Albizia adiantifolia Benth.
			Canarium schweinfurthii Engl.
			Berlinia giorgii Sol.
			Pentachletra eetveldeana
4	*Imbrasia epimethea*	Misa-misa (Kikongo/Kwilu) Essa (yansi/kwilu)	*Albizia adiantifolia* Benth.
			Pentaclethra eetveldeana De Wild. T. Durand.
			Canarium schweinfurthii Engl.
			Entandrophragma candolleii Harms.
			Entandrophragma angolense (Welw.) C. DC.
			Funtumia elastica (Preuss) Stapf.
			Macaranga monandra Thouars
			Ricinodendron heudelotii Mull.
5	*Imbrasia melanops /Imbrasia obscura*	Makangu (Kikongo/Kwilu) Miigway (Yansi/Kwilu)	*Aleurites mollucana* (L.) Wild.
			Macaranga monandra Thouars.
			Pentaclethra macrophylla Benth.
			Albizia adiantifolia Benth.
			Maesopsis eminii Engl.
6	*Anaphe sp.*	Miimfuun (Yansi/Kwilu) Nkanki (Kikongo/Kongo central)	*Bridelia feruginea* Benth.
			Lannea antiscorbutica A. Rich.
			Syzygium guineense (Wild.)
			Ceiba pentandra L.
7	*Antheua insignata*	Mindanda (Kikongo/Kwilu) Essiir (Y ansi/Kwilu)	*Hymenocardia ulmo'ides* Wall et Lindl.
			Macaranga monandra Thouars.
8	*Cymothoe caemis*	Mibamba(Kikongo/Kwilu) Ebamb (Yansi/Kwilu)	*Oncoba welwitschii* Forssk.
9	*Rhypoteryx poecilanthes*	Miilwoo (Yansi/Kwilu)	*Symphonia globulifera* L. f.
10	*Bunaopsis aurantiaca*	Sesenge (Mbole/Kisangani)	*Uapaca guineensis* Mull. Arg.
11	*Gonimbrasia hecate*	Afilika (Mbole/Kisangani)	*Albizia adiantifolia* Benth.
			Canarium schweinfurthii Engl.
			Macaranga monandra Thouars.
			Maesopsis eminii Engl.
			Uapaca guineensis Mull. Arg.
			Ricinodendron heudelotii Mull. Arg.
12	*Imbrasia oyemensis*	Tokala (Mbole/Kisangani)	*Entandrophragma angolense* (Welw.) C. DC.
			Entandrophragma candolleii Harms.
13	*Imbrasia truncata*	Makonzo (Lingala)	*Entandrophragma angolense* (Welw.) C. DC.
			Entandrophragma candolleii Harms.

14	*Pseudanthera discrepans*	Tombo (Mbole/Kisangani)	*Uapaca guineensis* Mull. Arg.
			Canarium schweinfurthii Engl.
			Maesopsis eminii Engl.

Analysis of Table 13 shows that 14 caterpillars will be able to nest on caterpillar plants planted in the SAF of the lbi-village. Most of these caterpillars are polyphagous and can feed on several plants and only three are monophagous, this is the case of *Cymothoe caemi* which only nests on the leaves of *Oncoba welwitschii, Rhypoteryx poecilanthes* which only feeds on the leaves of *Symphonia globulifera* and the *Bunaopsis aurantiaca* which only feeds on the leaves of *Uapaca guineensis* (LATHAM, 2008) and (OKANGOLA, 2007)

3.2. Discussion

Analysis of the host flora of edible caterpillars reveals 35 plant species belonging to 16 different families, grouped in 7 orders.

Among these families, the most characteristic are *Fabaceae* and *Euphorbiaceae* with 9 and 6 species respectively. The *Fabaceae* family is more diverse with 9 different species unlike KIDIKWADI (2012) who reported more species in this family.

The analysis of the density of the species shows that *Pentaclethra eedveldeana* has a high density with 39 individuals or 20.1% of the others. Contrary to the results of KIDIKWADI (2012) which shows the dominance of *Dialium englerianum* with 164 individuals in the Bombo-Lumene area.

The first position occupied by the *Fabaceae* family can be explained by the high number of individuals of these species in the environment studied, which is also the environment where the species are better adapted. The leaves are also rich in protein, which makes it easy for caterpillars to nest.

The analysis of the autoecological characteristics shows that the botanical inventory contains a large number of Guinean-Congolese species (54.4%). This can be explained by the fact that the Bateke plateau is in the phytochory of the regional centre of Guinean-Congolese endemism where the Lower Guinean-Congolese element dominates.

Furthermore, our analysis shows that Mesophanerophyte species are dominant at 85.7% of the flora studied. This result is contrary to that of KIDIKWAKI which shows the dominance of Microphanerophyte species at 71% in the Bombo-Lumene area.

From the diaspora point of view, sarcochorous species are numerous, i.e. 54.3% of the flora studied. This dominance is justified by the fact that the sarcochory is the mode of seed dispersal that characterizes many species of the evergreen rainforests (BOYEMBA, 2006 and LOMBE, 2007) where most of them are consumed by animals. Furthermore, our analysis shows that 91.4% of species are trees and very few shrubs. This is evidence that generally in the intertropical zone many of the phytophagous nest on trees than other morphological forms (BELESI, 2009).

In terms of height performance, we have 3 classes of species. The first class is that of the least performing species in height ranging from 0.2 to 1.3 m/year with 29 species, the second class is that of the species with average performance 1.4 to 2.5 m/year with 5 species and the third and last class is that of the most performing species ranging from 2.6 to 3.7 m/year with only one species.

As for the performance in diameter, we have three classes of species: the first one is the least performing species ranging from 0.4 to 1.9 cm/year with 25 species, the second class is the medium performing species ranging from 2.0 to 3.5 cm/year with 6 species and the last class is the most performing species ranging from 3.6 to 5.7 cm/year with 4 species.

These results on growth in height and diameter show that some species such as *Maesopsis eminii*, *Millettia drastica* and *Ceiba pentandra* are included in the class of the best performing species in height and diameter (2.6 to 3.7 m/year in height and 3,This coincides with the results of TANDJIEKPON and DAHDOVONON (2003) who measured the growth of *Acacia auriculiformis* with an annual performance of 1.83 m/year in height and 5.0 cm/year in diameter in the savannah region of Бёши. These results show us that we can classify our local species in the same performance class as Acacia and that we can use them well in our reforestation projects instead of Acacia.

This further proves that the phënology of most species is very complex and varied (BELESI, 1991).

Ethnozoological analysis has rëyë! ë that 3 species of caterpillars were observed in our site. These caterpillars are grouped into 2 major families. Notodontidae (2 species) and Nymphalidae (1 species).

Our results are inferior to those observed by N. LELEUP and H. DAEMS (1969) who identified 7 species in the Kwango savannah zone. DAEMS (1969) who identified 7 species in the Kwango savannah zone, inferior to the results of Malaise (1980) with 26 species identified in the Katanga province; to those of LATHAM (2000) 33 species in the Bas-Congo) and N'GASSE (2003) 24 species in the CAR region among which 13 species are in the Ngoto forest massif.

In our opinion, this difference in the results obtained would be justified, not only by the ecology of the habitats studied but also by their surface areas and types of substrates. Ours was sandy. Moreover, this observation is consistent with those of MALAISSE (1980), LATHAM (2001), MONZAMBE (2002), MOUSSA (2002), BALINGA (2003) and N'GASSE (2003) who underline the place of caterpillars in the diet of the populations of Central Africa and point out that the caterpillars consumed in this region of Africa belong to various families including Attacidae, Notodontidae and Nymphalidae

The results obtained in this ëtude show that most of the edible caterpillars have a polyphagous diet while the others are monophagous. This is the case of *Cymothoe caemi* which feeds only on the leaves of *Oncoba welwitschii.* It is an important resource kept in the sëtat to be consumed during dearth in northern Kwilu (BELESI, 2007).

According to studies conducted by MONZAMBE (2002), MOUSSA (2002), BALINGA (2003) and N'GASSE (2002), there are great differences in the feeding regime of caterpillars from one region to another. They cited the species *Bunaea alcinoe* which has *Uapaca guineensis* and *Musanga cecropioides* as host plants in the area of their studies while it also prefers *Persea americana, Mangifera indica* and *Dacryodes edulis* in central Kongo.

The analysis of our results shows that the caterpillar harvest period in the IBE zone is from February to December. These results are related to the results of N. LELEUP and H. DAEMS who reported the same harvest period in the Kwango region.

In our opinion, this period coincides with the off-season from June to August and the very wet period from September to December November, when the plants have leaves and can be eaten by caterpillars.

Furthermore, we note that *Imbrasia ertli,* which is the most common caterpillar species collected in our study area, was collected from May to June, unlike in Central Kongo where Latham (2008) reports its presence from October to December.

It is therefore important to note that caterpillar harvesting is a seasonal activity. The harvest period differs as reported by MALAISSE (1997) in Katanga, LATHAM (2000) and KANI-KANI (2000) in Central Kongo.

In our opinion, this discrepancy in the harvest period is explained by the difference in the study environment, the season and therefore the climatic factors.

Furthermore, our analysis also reveals that the majority of caterpillar host plants are used as medicinal plants (68.5%), 43% are used as food plants and as timber, 13 species or 37.1% as food, 25.7% as timber and 14.2% as firewood.

Our results are inferior to those found by OKANGOLA (2007) who found that in the Kisangani region the species were also 91.3% medicinal, 60.87% firewood, 47.83% fuelwood and 13.04% food.

Thus, medicinal plants are used to treat several diseases and are prepared in several forms (decoction, maceration, structuring and incineration). They are collected from the plant at several levels (root, leaves, latex, trunk or root bark) (WOME, 1983).

Finally, our analysis showed that most of the caterpillar host species planted in Ibi-village have several potential caterpillars that could nest there, including 14 potential edible caterpillars. This is the advantage of reforestation with local species initiated in this area.

CONCLUSION

The main objective of our study on the performance and potential of edible caterpillar host plants planted in Ibi-village FAS was to evaluate the performance and potential of edible caterpillar plants planted in Ibi-village agroforestry system. To do so, we used the observation method. At the end of our research, the following conclusions can be drawn:

The results obtained indicate a diversity of 35 species of edible caterpillar host plants, which are grouped into 7 orders and 16 families. This diversity confirms the first hypothesis of our work.

The species *Pentaclethra eedveldeana* is the best represented. On the other hand, the families *Fabaceae* and *Euphorbiaceae* contain more species than the other families.

Sarcochorous species are more dominant than the diaspora types identified.

The majority of the species are from the regional centre of endemism in Guinea and Congo. This justifies the belonging of our diction to this phytochory.

The most dominant leaf types are Mesophylls, which are mostly Mesophanerophytes, characteristic of most species in humid intertropical Africa.

In terms of height performance, we have 3 classes of species. The first class is that of the least performing species in height ranging from 0.2 to 1.3 m/year with 29 species, the second class is that of the species with average performance 1.4 to 2.5 m/year with 5 species and the third and last class is that of the most performing species ranging from 2.6 to 3.7 m/year with only one species.

As for the performance in diameter, we have three classes of species: the first one is the least performing species ranging from 0.4 to 1.9 cm/year with 25 species, the second class is the medium performing species ranging from 2.0 to 3.5 cm/year with 6 species and the last class is the most performing species ranging from 3.6 to 5.7 cm/year with 4 species.

Our results also allowed us to identify 3 species of edible caterpillars, grouped in 2 families: *Notodontidae* and Nymphalidae.

Most edible caterpillars have a polyphagous diet, unlike *Cymothoe caemi* which has a monophagous diet.

The harvest period runs from February to December and depends on the species.

Imbrasia ertli remains the most collected and most abundant caterpillar species in the area of Ibi-village, especially in the months of May and June 2020 and 2021, which have recently yielded a very large quantity of this caterpillar.

The edible caterpillar plants planted in Ibi-village have multiple uses and potentially 14 species of edible caterpillars will be able to nest on the species planted in Ibi-village and these results confirm all hypotheses.

This study has allowed us to highlight a significant diversity of edible caterpillar plants planted in Ibi-village, the performance of these species in height and diameter and also their potential. This study also showed us that we can put our local species in our reforestation projects instead of species such as *Acacia auriculiformis* because we have species with the same level of performance as the latter and they can be exploited in various forms and especially reforestation with edible caterpillar host plants can constitute an important source of income and animal protein.

In view of the above, we make the following recommendations:

- further studies to capitalise on these important resources for the country;
- the extension of the study to the entire Bateke plateau to clarify their timing;
- Encouraging awareness of the domestication of local plants, especially caterpillars, in order to perpetuate the species;
- the popularisation of caterpillar consumption, which contributes to the supply of protein and energy to the health system;
- raising awareness of the domestication of local plants, especially edible caterpillar host plants
- the involvement of the state in financing reforestation projects with local species.

BIBLIOGRAPHIC REFERENCES

1.	ACHOUNDONG, G., BONVALLOT, J. and HAPPI, Y. (2007): Le contact Foret-Savane dans l'Est du Cameroun et *Chromolaena odorata*: considërations preliminaries. Yaoundë, Cameroon ; Orstom, pp. 99-108.

2.	ACHOUR, A., AROUI, A., DEFAA, C., EL MOUSADIK, A. and MSANDA, F. (2011): Effect of dëfensing on floristic richness and densitë in two lowland argan groves. Actes du Premier Congres International de l'Arganier, Agadir 15-17, dëcembre 2011, 10p.

3.	ATANGANA, A. R, KHASA, D. P, CHANG, S.X, DEGRANDE, A. (2014): *Tropical agroforestry*. ISBN 978-94-007-7722-4, 467 springers.

4.	AUBREVILLE A. (1929) : Comment constituer une foret. Revue de Botanique appliquee, pp 560-568.

5.	AVENARD, J.-M. (1969): Rëflexюns on ГёШ research concerning the problems posed by forest-savanna contacts. Essai de mise au point et de bibliographie. ORSTOM, Paris, Ser. Initiations - Documentations techniques, N° 14, 162 p

6.	BADJI, M., SANOGO, D. and AKPO L., (2013): Effect of the age of dëfens on the reconstitution of woody vegetation in sylvopastoral areas of the southern groundnut basin (Senegal). Journal of Applied Biosciences 64 :4876 - 4887.

7.	BALINGA, M.P. (2003): *Caterpillars and edible larvae in the Cameroon forest zone*. Consultation report, FAO, Rome.

8.	BASAULA, N. (1989) : *Etude agrostologique et analyse fmanciere d'un projet bovin sur le plateau de Bateke* (Zaire), These de doctorat en Sciences Agronomiques, ULB, 181 p.

9.	BELESI, K.H. (2007): *Etude floristique, phytogeographique et phytosociologique des formations herbes du Kwilu septentrional*. Memoire DEA-ULB- Inedit. 87p.

10.	BELESI, K.H. (2009): *Etude floristique, phytogeographique et phytosociologique de la vegetation du bas-kasai* (Bandundu-RDC), Th. Doct. Inedit. 565p.

11.	BELESI, K.H. (2020): *Notes de Cours d'ecosystemes tropicaux*, L2 Environnement, UNIKIN, inedit, 47 p.

12.	BILOSO, M.A. (2008): *Valorisation des produits forestiers non ligneux des plateaux des Bateke en peripherie de Kinshasa (RD Congo)*, These de doctorat, Université Libre de Bruxelles, 252p.

13.	BILOSO, M.A. (2021): *Notes de cours de statistique appliquee a I'environnement,* L2 environnement, UNIKIN, unpublished.

14.	BOYEMBA, B. (2006): *Diversite et regeneration des essencesforestieres exploitees dans les forets des environs de Kisangani (RDC)* Mdmoire de DEA, ULB, Bruxelles, 101p.

15.	BURNEY, D.A. (1996): Climate change and fire ecology as factors in the Quaternary biogeography of Madagascar. pp. 49-58 in l.ourenco W.R., Biogdographie de Madagascar, Paris, Editions de l'ORSTOM.

16.   BURNEY, D.A. (1997): Theories and facts regarding Holocene environmental change before and after human colonization. P.75-89 in Patterson, B.D. and Goodman, S.M. Natural and Human-induced Change in Madagascar, Washington: Smithsonian Institution Press

17.   CAMARA R. (2000): The savannahs of the Dominican Republic and their exploitation by the "hatos" system 24p.

18.   CARTER, D.J. and HARGREAVES, B. and MINET, J. (1998): *Guide des chenilles d'Europe*. Delachaux et Nieste, Paris, 311 p.

19.   *Convention on Biological Diversity* (1992), 34 p.

20.   DE FOLIART (1992): *Insects as human food Crop protection* (11) 5: 395-399 p.

21.   DEFORESTA, H. (1990): Origine et évolution des savanes intermayombaises (R.P du Congo). II. Apport de la botanique forestiere. In: Lanfranchi and Schwartz eds. Quaternary Landscapes of Atlantic Central Africa. Orstrom, Paris pp. 326-355.

22.   DEGRANDE, A., ESSOMBA, H., BIKOUE, M.C. and KAMGA, A. (2007): *Domestication, Gender and Vulnerability, Participation of Women, Youth and the Poorest Categories in Agroforestry Tree Domestication in Cameroon*. World agroforestry Center, 85p.

23.   DUVIGNEAUD P. (1953): La flore et la vegetation du Congo meridional. Lejeunia, 16: 95- 124.

24.   FAO, NFP (2004): *Contribution of forest insects to food security*: The example of caterpillars in Central Africa. FAO, Rome, 140 p.

25.   GADE, D. (1996): Deforestation Madagascar. Human role in changing the face of the earth. Mountain Research and Development, pp. 101-116.

26.   *Intergovernmental Panel on Climate Change* (IPCC), (2003): Good practice recommendations for the land use, land use change and forestry sector, 594 p.

27.   HABARI, M., LEJOLY, J., and LUBINI, A. (2010): *Flora of the Pentaclethra eetveldeana community forests of the Kisantu region (DR Congo)*. In: X Van der Birgt, J. Van der Maesen and J.M. Onana (eds), Systematics and conservation of African plants pp. 643-651. Royal Botanic Garden, Kew.

28.   HIOL-HIOL, F., KEMEUZE V.A, KONSALA S., NJOUKAM R. (2013): Forest spaces of the savannahs and steppes of Central Africa: The forests of the Congo Basin, Etat des Forets, pp 165-171.

29.   HUMBERT, H. (1927): The destruction of an island flora by fire. Principaux aspects de la vëgëtation a Madagascar. Mëmoires de l'Acadëmie Malgache. Fasc. IV, 47 pp.

30.   IYONGO, L., WAYA, M., MARJOLEIN, V., De CANNIERE Ch., VERHEYEN E., DUDU AKAIBE B., ULYEL, J. and BOGAERT J. (2012): Anthropisation and edge effects: impacts on rodent diversity in the Reserve Forestière de Masako (Kisangani, R.D. Congo). Tropical Conservation Science Vol.5 (3) :270-283.

31.   JACQUIN, A. (2010): Dynamics of savanna vëgëtation in relation to 1 fire use in Madagascar. Analyse par sërie temporelle d'images de tëlëdëtection, Thëse de doctorat en Ecologie ; Universite de Toulouse, 146p.

32.   KANI-KANI, K. (2000): *Utilisations et gestion des ressources biologiques des forêts des communautes de la region de Kisantu (RDC)*. Memoire de DEA, Inedit, UNIKIN, 78p.

33.   KANKONDA, B. and WETSI, L. (1992): *Donnees preliminaires sur les chenilles de Kisangani et ses environs. Annales de lafaculte des sciences*, UNIKIS, vol8. Pp 113120.

34.   KAYUMBA, M., LUBINI, C., KIDIKWADI, E., HABARI, J.P. (2015): *Floristic study of the vegetation of the mature formation of the Bombo- Lumene estate and reserve (Kinshasa/RD Congo)*. International journal of Innovation and Applied studies, vol. 11, N°3 pp 716-727 p.

35.   KEAY, R.W.J. (1959): Derived savannah, derived from what? Bull. IFAN, ser. A (21), pp. 427-438.

36.   KIDIKWADI, E. (2012): *Estimation de Carbone sequestre par le peuplement vegetal a Dialium anglerianum et Hymenocardia acida dans le Domaine de chasse de Bombo-Lumene,* Plateau de Bateke/Kinshasa-RDC, 78 p.

37.   KISASA, R.K., PALATA, J.C., and PWEMA, V. (2005): *Observation on sexual behaviour of Giant Mole-Rat,* Gyptomys mechowi (Rodentia, Bathyerdae): Ethological an statistic approaches Researchgate 1-104 p.

38.   KLEIN, J. (2002): Deforestation in the Madagascar highlands. Established truth and scientific uncertainties. GeoJournal 56, p. 191-199.

39.   KOUBOUANA, F., NGOLIELE, A., NSONGOLA, G. (2007): *Evolution des parametres floristiques pendant la regeneration des forets de la reserve de la Lefini (Congo-Brazzaville),* annales de i'Universite Marien Ngouabi, Sciences et Techniques 8 (4) : 10-21 p.

40.   LARSEN, J. (1993): *Financial Mechanisms for Sustainable Conservation.* AFTES Working Paper No. 1 Environment and Sustainable Development Division, Technical Department, Africa Region, World Bank, Washington. 8 p.

41.   LATHAM, P. (2000): *Edible caterpillars and their food plants in the Bas-Congo province,* Myrstole publication, Centerbury, UK. 40 pp.

42.   LEJOLY, J. LISOWSKI, S. and NDJELE, M.B. (1988): *Vascular plants of Kisangani and Tsopo. Catalogue informatise de plantes.* Labo de botanique, systëmatique et ëcoiogie de

i'ULB, Brussels, 136 p.

43.	LELEUP, N. and DAEMS, H. (1969): *Les chenilles alimentaires du Kwango: Causes de leur rarefaction et mesures preconisees pour y remédier*, 24 p.

44.	LOMBE, B.L. (2007): *Contribution a l'etude de la phytodiversite de la reserve forestiere de Yoko (Ubundu, Republique Democratique du Congo)* Mëmoire de DES incdit, Faculte des Sciences, UNIKIS, 72 p.

45.	MALAISSE, F. (1997): *Se nourrir en foret claire africaine.* Approche ecologique et nutritionnelle. Presse agronomique de Gembloux CTA, Wageningen, Belgium, 384 p.

46.	MALDAGUE, M. (2010): *Treatise on Tropical Environmental Management*, Integrated Development of Tropical Regions, Systemic Approach. Notions- Concepts-Methodes, Tome I, 441 p.

47.	MALELE. M., (2003) : Situation des ressources genetiques forestiëres de la Republique democratique du Congo, division des ressources forestieres, working paper FGR/56F, FAO, Rome, Italy, 48p.

48.	MALOBA, J.D., (2011): Forest-savanna mosaic and exploitation of forest resources in Gabon, Geo-Eco-Trop 35: 41 - 50.

49.	MONZAMBE, M. (2002): *Contribution a l'exploitation des chenilles et autres larves comestibles dans la lutte contre l'insecurite alimentaire et la pauvreté en Republique Democratique du Congo (RDC).* Consultation report, FAO, Rome. Pp 35-54.

50.	MUHASHY, H.F., NLANDU, L. and MALIO, N. (2011): *Habitats of the reserve Domaine de chasse de Bombo-Lumene (RD Congo)*, Kiteke lexicon of plants observed in these environments. Brussels, 114 p.

51.	N'Gasse, G. (2003): *Contribution of edible caterpillars/larvae to the reduction of food insecurity in Central African Republic (CAR).* Consultation report, FAO, Rome.

52.	PARENT, S. (1990): *Dictionnaire des sciences de l'Environnement*, 748 p.

53.	PERRIER, H. (1921): La vegetation malgache. P. 12-26, Annales de l'Institut Botanico-geologique Colonial de Marseille, Ser. 3(9), 268 p.

54.	RAKOTOARIMANANA, J.R., GONDARD, V., RANAIVOARIVELO, H., and CARRIERE, S. (2008): Influence of grazing on floristic diversity, production and forage quality of a savanna in the Malagasy Highlands (Fianarantsoa region). Secheresse, vol. 19, no. 1, p. 39-46.

55.	RAVEN, P.H., BERG, L.R., and HASSENZAHI, D.M. (2009): *Environment, Edition DE Boeck universite*, Rue des Minimes, 39 B-1000 Brussels, 687 p.

56.	*Republique Democratique du Congo*, Code forestier (2002) du journal officiel-Numero Special 6 novembre, 102 p.

57.   SCHERECKENBERG, K., AWONO, A., MBOSSO, C., NDOYE, O. & TCHOUNDJEU, Z. (2006): *Domesticating indigenous fruit trees as a contribution to paverty reduction forest,* Tress and Livelihoods Vol.16, pp 35-51.

58.   SMART, P. (1981): *Encyclopedie des papillons du monde entier, plus de 2000 especes en couleur et grandeur vie.* Edition Bordas Nature, Paris.

59.   TAFOKOU JIOFACK, R.B. (2014): Population management of a multiple-use non-timber forest product: *Tetracarpidium conophorum* (Mull. Arg.) Hutch & Dalz (Euphorbiaceae) in forest management systems in Cameroon. D. thesis, ERAIFT, 285 p.

60.   TANDJIEKPON A. and DOVONON J. (2003): Influence of seed provenance on forest plant development: the case of *Acacia auriculiformis* A. CUNN. EX BENTH. Word forestery congres, Quebec city, Canada, 15p.

61.   VANDE, J-P. (2004): Foret d'Afrique centrale: la nature et l'homme, Editions Iannoo SA. Belgium, 357 p.

62.   VERMEULEN, C. and LANATA, F. (2006): Le domaine de chasse de Bombo-Lumene : un espace naturel en përΠ aux frontieres de Kinshasa, pares et reserves, Vol. 61 N°2 pp 4-8.

63.   WIERSUM, K.F, (1996): Domestication of valuable trees species in agroforestry systems, evolutionary stages from gathing to breeding. In: Leakey, R.R. B, Temu, A.B. and Melnyk, M(eds). Domestication and commercializationof non-timber forest products in agroforestry systems. Technical papers, Non-Wood forest products 9. Fao, Rome, Italy. 147-158 pp.

64.   WOMBE, B. (1985): Recherches ethnopharmacognosiques sur les plantes mëdicinales utilisëes en mëdecine traditionnelle a Kisangani (Haut-Zaire) These indite, T2 ? Facu№ des sciences, ULB, Belgium. 562 p.

65.   YOKA, J., LOUMETO, J.J. and VOUIDIBIO, J. (2007): Quelques cara^ristiques ëcologiques des savanes de la zone d'Ollombo (cuvette Congolaise, Rëpublique du Congo), Facu№ des Sciences. Annales de i'Universile Marien Ngouabi, Sciences et Techniques, 8 (4): 74- 87

66.   YOUNG, A. (1995): Agroforestry for Soil Conservation, Technical Centre for Agricultural and Rural Co-opëration, Brussels, 183 p.

SITE CONSULTED

1. Http: //www. statistics-world.com/congo_kinshasa.htm, accessed 22.07.2020

I want morebooks!

Buy your books fast and straightforward online - at one of world's fastest growing online book stores! Environmentally sound due to Print-on-Demand technologies.

Buy your books online at
www.morebooks.shop

Kaufen Sie Ihre Bücher schnell und unkompliziert online – auf einer der am schnellsten wachsenden Buchhandelsplattformen weltweit! Dank Print-On-Demand umwelt- und ressourcenschonend produzi ert.

Bücher schneller online kaufen
www.morebooks.shop

KS OmniScriptum Publishing
Brivibas gatve 197
LV-1039 Riga, Latvia
Telefax: +371 686 204 55

info@omniscriptum.com
www.omniscriptum.com

MIX
Papier aus verantwortungsvollen Quellen
Paper from responsible sources
FSC® C105338

Printed by Books on Demand GmbH, Norderstedt / Germany